A. Arockia Bazil Raj

Free Space Optical Communication

Also of Interest

Modern Communications Technology
Natasa Zivic, Miroslav Dukic, 2016
ISBN 978-3-11-041337-3, e-ISBN 978-3-11-041338-0,
e-ISBN (EPUB) 978-3-11-042390-7

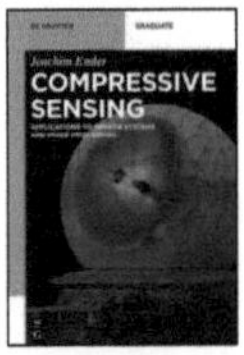

Compressive Sensing
Joachim Ender, 2016
ISBN 978-3-11-033531-6, e-ISBN 978-3-11-033539-2,
e-ISBN (EPUB) 978-3-11-039027-8

Signals and Systems
Gang Li, Liping Chang, Shen Li, 2015
ISBN 978-3-11-037811-5, e-ISBN 978-3-11-037954-9,
e-ISBN (EPUB) 978-3-11-041684-8

Advanced Optical Technologies
Michael Pfeffer (Editor in Chief), 6 issues/year
ISSN 2192-8584

Journal of Optical Communications
Ralph Th. Kersten (Editor in Chief), 4 issues/year
ISSN 2191-6322

A. Arockia Bazil Raj

Free Space Optical Communication

System Design, Modeling, Characterization, and Dealing with Turbulence

DE GRUYTER
OLDENBOURG

Dr. A. Arockia Bazil Raj
Associate Professor
Laser Communication Laboratory (LCL) Facility
Electronics and Communication Engineering
Kings College of Engineering
Punalkulam – 613 303
Tamil Nadu, India
E-Mail: brazilraj.a@kingsindia.net

ISBN 978-3-11-044995-2
e-ISBN (PDF) 978-3-11-045261-7
e-ISBN (EPUB) 978-3-11-045016-3
Set-ISBN 978-3-11-045262-4

Library of Congress Cataloging-in-Publication Data
A CIP catalog record for this book has been applied for at the Library of Congress.

Bibliographic information published by the Deutsche Nationalbibliothek
The Deutsche Nationalbibliothek lists this publication in the Deutsche Nationalbibliografie; detailed bibliographic data are available on the Internet at http://dnb.dnb.de.

Cover image: IdealPhoto30/iStock/Thinkstock
Typesetting: PTP-Berlin, Protago-TEX-Production GmbH, Berlin
Printing and binding: CPI books GmbH, Leck
♾ Printed on acid-free paper
Printed in Germany

www.degruyter.com

Preface

Free Space Optical Communication (FSOC) is an effective alternative technology to meet the Next Generation Network (NGN) demands as well as highly secured (military) communications. FSOC includes various advantages like last mile access, easy installation, free of Electro Magnetic Interference (EMI)/Electro Magnetic Compatibility (EMC) and license free access etc. In FSOC, the optical beam propagation in the turbulent atmosphere is severely affected by various factors suspended in the channel, geographical location of the installation site, terrain type and meteorological changes. Therefore a rigorous experimental study over a longer period becomes significant to analyze the quality and reliability of the FSOC channel and the maximum data rate that the system can operate since data transmission is completely season dependent.

The FSOC transmitter and receiver experimental setup is established for a link range of 0.5 km at an altitude of 15.25 m. A low cost and highly accurate weather station is developed in the year 2009 and used to continuously collect the data corresponding to the meteorological parameters' fluctuations. All the collected data are stored in data logging computer. The measurement accuracy of the weather station is calibrated against the standard instrument and average uncertainty of 1.1 is achieved in all the measurements. The atmospheric attenuation and turbulence (C_n^2) are measured using the constructed optoelectronic assembly (scintillometer) setup. New models are developed based on the experimental data collected at the experimental site to predict the atmospheric attenuation and turbulence strength (C_n^2) as a function of local meteorological parameters. The prediction accuracy and suitability of the new models developed at our test field are validated against the prediction results of selected existing models and also with measurement data. A coefficient of determination (R^2) of prediction of 98.76 % for atmospheric attenuation and of 98.93 % for turbulence strength (C_n^2) were achieved.

The power level, beam divergence, scintillation and beam wandering etc. of the received optical wave on the detector plane is continuously monitored for several days during different seasons. Response Surface Model (RSM) and Artificial Neural Network (ANN) based neural-controllers are developed using the behavior of the channel and information on beam position. These two controllers are developed in MATLAB environment and their performance is studied in open loop, closed loop, narrow band and wide band disturbances and x, y, and z direction tilt. It is found that the neural-controller exactly generates the correction control signal as required to stabilize the beam at the detector plane rather than the RSM controller. A novel pipelined-parallel digital architecture is designed to implement the neural controller in Field Programmable Gate Array (FPGA) and then high speed parallel-pipelined control action is performed. The maximum control signal percentage of prediction error of 2.21 % is observed.

A simplex FSOC optoelectronics assembly with beam Aligning, Positioning and Tracking (APT) is established to couple the Power In the Bucket (PIB) to the detector plane. The power fluctuation due to the beam wandering is controlled and brought down to 0.312 V with the APT system. The optical beam is modulated at the Asynchronous Transfer Mode (ATM) rate of 155 Mbps using an On-Off-Keying Intensity Modulation Direct Detection Non-Return-to-Zero (OOK-IM-DD-NRZ) modulation scheme. The quantitative analysis of FSOC data link is analyzed in terms of Signal to Noise Ratio (SNR) i.e. the Q-factor, link margin and Bit Error Rate (BER) as a function of turbulence strength (C_n^2). The significant improvements of data link reliability with and without the Aligning Positioning Tracking (APT) system are investigated. Finally, the minimum BER of $6.4 \cdot 10^{-9}$ is maintained for different outdoor environments.

All the original contributions, newness, findings and experimental results etc. are reported in this book.

I am very grateful to Dr. U. Sripati Acharya, Professor, National Institute of Technology, Surathkal, Karanataka, India, Dr. Chris Dainty, Professor, National University of Ireland, Galway, Ireland, and Dr. J. P. Lancelot, Scientist-'E', Indian Institute of Astrophysics, Bangalore, Karanataka, India, for their expert review and comments that scientifically strengthened the materials published inside the book.

Thanjavur (India), November 2015 Dr. A. Arockia Bazil Raj

To my Mother **A. Jothy Mary** and my Father **A. Anthoni Samy,**

To my Wife **A. Johnsi Rosita** and my Daughter **A. Feona Hydee Mitchell,** and

To **Mr. J. Niranjan Samuel**, who has admirably and unreservedly extended his helping hand and support throughout this book.

Contents

Preface — V
List of Tables — XIII
List of Figures — XV
List of symbols and abbreviations — XXIII

1 **Introduction — 1**
1.1 Background — **1**
1.2 Research motivation — **2**
1.3 Characteristics of FSOC — **5**
1.3.1 Directionality of the light beam — **5**
1.3.2 Form factors, i.e. size and power per bit — **5**
1.3.3 Ability to be operated license-free worldwide and quick installation — **6**
1.3.4 Wavelength selection criteria — **6**
1.3.5 Challenges and limitations — **7**
1.4 Applications and advantages — **7**
1.5 Research objectives — **8**
1.6 Original contributions – newness and achievements — **8**
1.7 Thesis organization — **9**

2 **Real-time measurement of meteorological parameters for estimating low altitude atmospheric turbulence strength (C_n^2) — 13**
2.1 Introduction — **13**
2.2 Background and related works — **14**
2.3 Field test experimental setup and measurement protocol — **16**
2.4 Sensor interfacing architectures and data acquisition protocols — **18**
2.4.1 Wind speed measurement – cup anemometer — **19**
2.4.2 Relative humidity and temperature measurement – SHT11 Sensor — **21**
2.4.3 Absolute pressure measurement – SCP1000-D01 sensor — **24**
2.5 Communication protocol and frame format — **29**
2.6 Performance calibration of the proposed measurement system — **31**
2.7 Atmospheric turbulence strength (C_n^2) estimation — **34**
2.8 Experimental results and discussions — **36**
2.8.1 Data for 28th December 2012, winter — **37**
2.8.2 Data for 5th March 2013, presummer — **39**
2.8.3 Data for 17th May 2013, summer — **39**
2.8.4 Data for 13th June 2013, monsoon — **41**
2.8.5 Data for 16th November 2013, rainy — **43**

2.9 Advantages — 44
2.10 Summary — 44

3 Comparison of different models for ground-level atmospheric attenuation and turbulence strength (C_n^2) prediction with new models according to local weather data for FSO applications — 47
3.1 Introduction — 47
3.2 Background and related works — 51
3.3 Experimental setup and description of optoelectronic assembly — 53
3.4 Comparison models of atmospheric influence on optical propagation — 54
3.4.1 Atmospheric attenuation — 55
3.4.2 Atmospheric optical turbulence strength — 58
3.5 Formulation of the mathematical model — 60
3.5.1 Atmospheric attenuation — 60
3.5.2 Atmospheric turbulence strength (C_n^2) from meteorological measurements — 63
3.6 Experimental results and data analysis — 67
3.6.1 Comparison of the predicted attenuation data with measured values — 67
3.6.2 Comparison of predicted C_n^2 data with measured values — 80
3.7 Summary — 92

4 Mitigation of beam wandering due to atmospheric turbulence and prediction of control quality using intelligent decision making tools — 93
4.1 Introduction — 93
4.2 Background and related works — 95
4.3 FSO link – optoelectronic assembly and setup description — 97
4.4 Steady state response analysis — 100
4.4.1 Optoelectronic position detector — 100
4.4.2 Piezo driving amplifier — 101
4.4.3 Piezoelectric actuators — 101
4.5 Development of response surface models — 103
4.6 Development of the neural network model — 105
4.7 Experimental results and discussion — 109
4.7.1 Verification and validation of RSM and neural-controller model — 109
4.7.2 Behavioral study of neural-controller in beam alignment — 113
4.7.3 Analysis of receiver signal quality improvement — 114
4.8 Summary — 115

5 **Low power and compact RSM and neural-controller design for beam wandering mitigation with a horizontal-path propagating Gaussian-beam wave: focused beam case — 117**
5.1 Introduction — **117**
5.2 Background and related works — **119**
5.3 Experimental plant configuration and centroid error computation — **120**
5.4 Formulation and implementation of direct controller — **122**
5.5 Hardware architecture and implementation of the neural-controller — **124**
5.5.1 Clock manager unit — **128**
5.5.2 Signal digitization and data preprocessing unit — **128**
5.5.3 Weight and bias memory management circuit — **130**
5.5.4 Neuron unit — **131**
5.5.5 Data Routing Ring Circuit (DRRC) — **132**
5.5.6 Multiply-Accumulator unit (MAC) — **134**
5.5.7 Serial communication manager — **135**
5.6 Experimental results and discussion — **136**
5.6.1 Control schemes validation and evaluation in open loop decision-making — **136**
5.6.2 Performance study of closed loop experiment with intensity feedback control — **140**
5.6.3 Analysis of beam spot auto alignment and reduction of wandering — **143**
5.6.4 Behavioral study of effective scintillation index and impulse response — **144**
5.7 Summary — **146**

6 **Quality metrics and reliability analysis of ground-to-ground free space laser communication in different weather conditions together with beam steering system — 149**
6.1 Introduction — **149**
6.2 Background and related works — **154**
6.3 Theory and numerical technique for channel effect and BER evaluation — **156**
6.4 Simplex data transmission experimental setup and its description — **159**
6.5 Experimental results and data analysis — **161**
6.5.1 Comparative evaluation of received signal statistics — **162**
6.5.2 Impact validation of beam wandering compensation system — **166**

6.5.3 Quantitative analysis of atmospheric turbulence effects on communication parameters – improvement and reliability — **170**
6.6 Summary — **173**

7 Conclusions and future work — 175
7.1 Conclusions — **175**
7.2 Future work — **178**

References — 183
Index — 201

List of Tables

Tab. 2.1 Parameters of the optical link. —— **17**
Tab. 2.2 A portion of measured meteorological data recorded in an Excel work sheet. —— **37**
Tab. 2.3 A portion of a daywise local meteorological data and estimated optical turbulence strength record. —— **37**

Tab. 3.1 Experiment parameters of the optical link. —— **55**
Tab. 3.2 R^2 value of Response Surface Model (RSM) for attenuation and turbulence strength prediction. —— **62**
Tab. 3.3 Percentage deviations between experimental and prediction values of the proposed models for attenuation and turbulence strength respectively. —— **66**

Tab. 4.1 Experimental design and their responses (observed values). —— **104**
Tab. 4.2 R^2 values of response surface models and neural-controller for C_x and C_y. —— **110**
Tab. 4.3 Results from confirmatory experiment ($V_{Ref} = 10, 5, 8$ V). —— **110**
Tab. 4.4 Validation test results of developed response surface model (full) and neural-controller for C_x and C_y. —— **112**

Tab. 5.1 Example of a process-time diagram (brief) showing the pipeline of the computation process of the neural-controller with process description and time. —— **126**
Tab. 5.2 Validation test results of developed response surface model (full) and neural-controller for C_x and C_y. —— **138**
Tab. 5.3 Empirical standard deviations and position magnitude average of experiments using the population length of 10 000 data points in open and closed loop control configuration. —— **141**

Tab. 6.1 Comparison between RF and FSOC communication systems. —— **150**
Tab. 6.2 BER summary of FSOC data transmission experiments and results. —— **162**
Tab. 6.3 Experimentally measured radial displacement and BER statistics. —— **169**

List of Figures

Fig. 1.1 A typical concept of an FSOC network. —— **3**

Fig. 1.2 Contribution of Rayleigh scattering, Mie scattering and water vapor to atmospheric transmittance. —— **6**

Fig. 1.3 Outline of the research roadmap and original contribution. —— **11**

Fig. 2.1 Optical propagation path for the local C_n^2 field data measurement and acquisition with direct transmission test equipment (scintillometer setups). The receiver system is seen on the building at the left and the transmitter system is seen on the tower at the right. —— **16**

Fig. 2.2 Photograph of sensor compartment – sensors are protected from the direct influence of environmental changes and kept in a waterproof housing. —— **19**

Fig. 2.3 Components assembly and design structure of cup anemometer – kept in a waterproof cabin. —— **20**

Fig. 2.4 Circuit schematic and architecture of cup anemometer configuration for wind speed measurement. —— **20**

Fig. 2.5 VHDL pseudocode of wind speed measurement. —— **21**

Fig. 2.6 Wind speed measurement simulated timing diagram – Modelsim results for the value $(000000010000)_2 = (16)_{10} = 2.75\,\mathrm{m\,s^{-1}}$. —— **21**

Fig. 2.7 Digital architecture for temperature and relative humidity measurement-TWI interfacing protocol. —— **22**

Fig. 2.8 Finite state machine control engine state transition flow for temperature and relative humidity measurement. —— **23**

Fig. 2.9 VHDL pseudocode of RH and T measurement. —— **25**

Fig. 2.10 Timing diagram of T and RH measurement simulated in Modelsim for the values $01100100010100 = (6420)_{10} = 24.10\,°\mathrm{C}$ and $010101111000 = (1400)_{10} = 46.09\,\%$. —— **25**

Fig. 2.11 Digital architecture of pressure measurement: SPI communication interface protocol. —— **26**

Fig. 2.12 Finite state machine control engine state transition flow for pressure measurement in triggered mode. —— **27**

Fig. 2.13 VHDL pseudocode of P measurement. —— **28**

Fig. 2.14 Timing diagram of P measurement simulated in Modelsim for the value $(1100010111000010000)_2 = (405008)_{10} = 101.252\,\mathrm{kPa}$. —— **29**

Fig. 2.15 (a) UART–RS232 communication protocol architecture and (b) communication frame format. —— **30**

Fig. 2.16 Timing diagram of UART communication protocol simulated in Modelsim for $Ws = 00001001111000 = (632)_{10} = 98.2458\,\mathrm{m\,s^{-1}}$, $T = 1101000001110 = (6670)_{10} = 26.6\,°\mathrm{C}$, $RH = 010010101001 = (1193)_{10} = 39.63\,\%$, and $P = 1100100000110010000 = (410000)_{10} = 102.5\,\mathrm{kPa}$. —— **30**

Fig. 2.17 Comparison between the standard and proposed measurement data for ascending and descending input variations: standard (blue), test (red), linear regression (black) and residual (green). —— **32**

Fig. 2.18 Performance plots of coefficient of determination (red): (a) wind velocity, (b) relative humidity, (c) temperature, and (d) pressure. The inset figures (black) show the achieved accuracy of measurement uncertainty (U_e) against the respective normalized environmental parameter. —— **33**

Fig. 2.19 Diurnal time series profile of meteorological parameters over a one day period. (a–d) A comparison of time series plots of atmospheric turbulence strength (C_n^2) estimated using PAMELA model (blue) and direct transmission measurement using scintillometer setup, i.e. based on signal intensity ensemble average data (red). (e) Correlation statistic plot for measured and estimated C_n^2 values. (f) MN: midnight and Nn: noon. —— **38**

Fig. 2.20 Diurnal time series profile of meteorological parameters over a one day period. (a)–(d) A comparison of time series plots of atmospheric turbulence strength (C_n^2) estimated using PAMELA model (blue) and direct transmission measurement using scintillometer setup, i.e. based on signal intensity ensemble average data (red). (e) Correlation statistic plot for measured and estimated C_n^2 values. (f) MN: midnight and Nn: noon. —— **40**

Fig. 2.21 Diurnal time series profile of meteorological parameters over a one day period. (a–d) A comparison of time series plots of atmospheric turbulence strength (C_n^2) estimated using PAMELA model (blue) and direct transmission measurement using scintillometer setup, i.e. based on signal intensity ensemble average data (red). (e) Correlation statistic plot for measured and estimated C_n^2 values. (f) MN: midnight and Nn: noon. —— **41**

Fig. 2.22 Diurnal time series profile of meteorological parameters over a one day period. (a–d) A comparison of time series plots of atmospheric turbulence strength (C_n^2) estimated using PAMELA model (blue) and direct transmission measurement using scintillometer setup, i.e. based on signal intensity ensemble average data (red). (e) Correlation statistic plot for measured and estimated C_n^2 values. (f) MN: midnight and Nn: noon. —— **42**

Fig. 2.23 Diurnal time series profile of meteorological parameters over a one day period. (a–d) A comparison of time series plots of atmospheric turbulence strength (C_n^2) estimated using PAMELA model (blue) and direct transmission measurement using scintillometer setup, i.e. based on signal intensity ensemble average data (red). (e) Correlation statistic plot for measured and estimated C_n^2 values. (f) MN: midnight and Nn: noon. —— **43**

Fig. 3.1 Bird's eye view (GPS) of optical propagation path for atmospheric attenuation and C_n^2 field data measurement. The receiver system is located on the building at the left (marked by red balloon) and the transmitter system is located on the tower at the right (marked by blue balloon). —— **48**

Fig. 3.2 Schematic diagram of Laser Communication Laboratory (LCL) experimental setup (transmitter and receiver) constructed to measure atmospheric attenuation and turbulence strength and modeling. —— **53**

Fig. 3.3 Snapshot showing the optoelectronic components assembly on the vibration damped optical breadboard (a) transmitter and (b) receiver. —— **54**

Fig. 3.4 Illustrations of input-output response surface plots. (a) attenuation vs wind speed and temperature, (b) attenuation vs wind speed and relative humidity, and (c) attenuation vs temperature and relative humidity —— **62**

Fig. 3.5 Residual plots of the developed regression model (cubic equation) for optical attenuation. —— **63**

Fig. 3.6 Illustrations of input-output response surface plots. (a) turbulence strength (C_n^2) vs wind speed and temperature, (b) C_n^2 vs wind speed and relative humidity, and (c) C_n^2 vs temperature and relative humidity. Note that the multiplication factor in the z-axis is 10^{-14}. — **65**
Fig. 3.7 Residual plots of developed regression model (model equation V) for turbulence strength. — **66**
Fig. 3.8 Diurnal time series profile of measured macrometeorological parameters (a–c) and visibility (d). MN: midnight and Nn: noon. — **68**
Fig. 3.9 Graphical comparison of time series profile of atmospheric attenuation predicted using selected existing models and proposed models and measured from direct beam transmission signal power (a). Root mean square error value against the particular model (b). — **69**
Fig. 3.10 Diurnal time series profile of measured macrometeorological parameters (a–c) and visibility (d). MN: midnight and Nn: noon. — **71**
Fig. 3.11 Graphical comparison of time series profile of atmospheric attenuation predicted using selected existing models and proposed models and measured from direct beam transmission signal power (a). Root mean square error value against the particular model (b). — **72**
Fig. 3.12 Diurnal time series profile of measured macrometeorological parameters (a–c) and visibility (d). MN: midnight and Nn: noon. — **74**
Fig. 3.13 Graphical comparison of time series profile of atmospheric attenuation predicted using selected existing models and proposed models and measured from direct beam transmission signal power (a). Root mean square error value against the particular model (b). — **75**
Fig. 3.14 Diurnal time series profile of measured macrometeorological parameters (a–c) and visibility (d). MN: midnight and Nn: noon. — **76**
Fig. 3.15 Graphical comparison of time series profile of atmospheric attenuation predicted using selected existing models and proposed models and measured from direct beam transmission signal power (a). Root mean square error value against the particular model (b). — **77**
Fig. 3.16 Diurnal time series profile of measured macrometeorological parameters (a–c) and visibility (d). MN: midnight and Nn: noon. — **78**
Fig. 3.17 Graphical comparison of time series profile of atmospheric attenuation predicted using selected existing models and proposed models and measured from direct beam transmission signal power (a). Root mean square error value against the particular model (b). — **79**
Fig. 3.18 Diurnal time series profile of meteorological parameters: wind speed, temperature, relative humidity and barometric pressure (a–d). — **81**
Fig. 3.19 Graphical comparison of time series profile of atmospheric turbulence strength (C_n^2) predicted using selected existing models and proposed models and measured from direct beam transmission system (a). Sum of absolute error value against the particular model (b). — **82**
Fig. 3.20 Diurnal time series profile of meteorological parameters: wind speed, temperature, relative humidity and barometric pressure (a–d). — **83**
Fig. 3.21 Graphical comparison of time series profile of atmospheric turbulence strength (C_n^2) predicted using selected existing models and proposed models and measured from direct beam transmission system (a). Sum of absolute error value against the particular model (b). — **84**

Fig. 3.22 Diurnal time series profile of meteorological parameters: wind speed, temperature, relative humidity and barometric pressure (a–d). — **86**

Fig. 3.23 Graphical comparison of time series profile of atmospheric turbulence strength (C_n^2) predicted using selected existing models and proposed models and measured from direct beam transmission system (a). Sum of absolute error value against the particular model (b). — **87**

Fig. 3.24 Diurnal time series profile of meteorological parameters: wind speed, temperature, relative humidity and barometric pressure (a–d). — **88**

Fig. 3.25 Graphical comparison of time series profile of atmospheric turbulence strength (C_n^2) predicted using selected existing models and proposed models and measured from direct beam transmission system (a). Sum of absolute error value against the particular model (b). — **89**

Fig. 3.26 Diurnal time series profile of meteorological parameters: wind speed, temperature, relative humidity and barometric pressure (a–d). — **90**

Fig. 3.27 Graphical comparison of time series profile of atmospheric turbulence strength (C_n^2) predicted using selected existing models and proposed models and measured from direct beam transmission system (a). Sum of absolute error value against the particular model (b). — **91**

Fig. 4.1 Photograph of Laser Communication Laboratory (LCL) facility: FSOL receiver (left) and transmitter (right) laboratories developed at information technology block and tower constructed for this work respectively. — **97**

Fig. 4.2 Schematic of FSOL experimental setup for 0.5 km optical link. Left-hand side: receiver with beam steering optoelectronic assembly and right-hand side: transmitter. — **98**

Fig. 4.3 Beam spot of 4 mm diameter (a) centered (0, 0) mm and (b) displaced (X_{dist}, Y_{dist}) mm beam spot on the optoelectronic position detector surface. — **99**

Fig. 4.4 Sensitivity measurement and calibration of OPD-PTQ100 for the beam position drift from center left to center right on the x-axis (V_{Ex}) and from center top to center bottom on the y-axis (V_{Ey}) with curve-fit model. — **100**

Fig. 4.5 Open loop sensitivity measurement and calibration of piezo amplifier for (a) x-channel (b) y-channel with curve-fit models and piezo platform for (c) x-channel from left-hand side to right-hand side and (d) y-channel from top to bottom. — **102**

Fig. 4.6 One pattern of measured hysteresis loop obtained for cyclic variation of control signal in process nonlinearity testing. — **103**

Fig. 4.7 Proposed structure of neural-controller with 2–12–9–2 multilayer perceptron model. — **106**

Fig. 4.8 Error rate for the learning function (error rate versus iteration). The learning procedure is stopped when the final error is below the goal. — **108**

Fig. 4.9 One pattern of neural network structure with coded weights and bias values obtained from back-propagation training algorithm. — **109**

Fig. 4.10 Normal probability plot of considered full model: percent versus residual for x- (left) and y-channel (right) control. — **111**

Fig. 4.11 Normal probability plot of proposed neural-controller: percent versus residual for x- (left) and y-channel (right) control. — **111**

Fig. 4.12 Open loop plant response for multiple setpoint errors: Reference control voltages are (a) $0.2157 \rightarrow 3.8 \rightarrow -2.3 \rightarrow 1.6$ V for the x-channel and (b) $-0.1634 \rightarrow 2 \rightarrow 5.3 \rightarrow -1.6$ V for the y-channel. — **112**

Fig. 4.13 Percentage of error of control signal predicted by the neural-controller in open loop configuration for x- and y-channels. — **113**

Fig. 4.14 Figures on the left show compass plot of time series laser beam spot centroid motion on OPD with beam steering control off and on conditions. Figures on the right show the corresponding histogram of radial distance (from plane center to beam centroid). — **114**

Fig. 4.15 Figures on the left show the time series of photodiode output without (top) and with (bottom) beam steering and on the right show the corresponding histogram. — **115**

Fig. 5.1 Optical propagation path: The receiver system is located on the IT block at the left and the transmitter system is located on a tower at the right. — **120**

Fig. 5.2 Schematic of the free space optical communication experimental setup for 0.5 km data link. Left-hand side: receiver with beam steering optoelectronic assembly and right-hand side: transmitter (red arrow line: optical path and black arrow line: signal path). — **121**

Fig. 5.3 Hardware design flow graph of data normalization algorithm in PVs preprocessor unit. Error signal V_{Ex} and V_{Ey} variations −10 V to +10 V are normalized with respect to V_{Ref}. Error signal of 0 V is equal to 800H = $(2048)_{10}$. — **123**

Fig. 5.4 Direct controller design based on the response surface model with pipeline registers. — **124**

Fig. 5.5 (a) Typical back-propagation neural network trained pattern coded with weights and bias values, Standalone Linear Time Invariant (LTI) system with neural-controller in (b) open loop: incident light falls on the OPD and FSM. FSM driven signals are approximately proportional to the values of process variables, and (c) closed loop: incident light falls on FSM. The reflected light falls on the OPD and a control signal is applied in feedback control configuration. — **125**

Fig. 5.6 Global block diagram of clock manager, data acquisition, neural-controller and UART implementation in the FPGA device. A, B, C, D, E, F, and G are main subcircuits; a, b, c, d, e, f, and g are pipeline stages. — **127**

Fig. 5.7 (a) Schematic diagram of digital clock manager unit and (b) eight channel and 12 bit parallel out bidirectional (−10 V to +10 V) A/D interfacing and data acquisition circuit. — **129**

Fig. 5.8 Synaptic weight and bias value memory organization in single precision (32 bit) floating point data format in RAM for (a) hidden layer 1 (HL1), (b) output layer (OL), and (c) hidden layer 2 (HL2). — **130**

Fig. 5.9 (a) Hardware design flow graph of hyperbolic tangent sigmoidal (bipolar) activation function model as the second order nonlinear function and (b) plot of real and modeled neuron behavior with $\sigma = 0.032$, $\theta = 1$, $\beta = 255$, $n \in \{-255, 255\}$, and $L = \pm\beta/2$. — **132**

Fig. 5.10 Neuron output data routing ring circuits and serial in and serial out circular shift register array to accomplish the multiply-accumulation computation for (a) Hidden Layer 2 (HL2) and (b) Output Layer (OL). — **133**

Fig. 5.11 Multiply-accumulation architecture implementation design flow graph. — **134**

Fig. 5.12 (a) UART RS232 communication protocol architecture and (b) communication frame format. — **135**

Fig. 5.13 Time series plot of performance of open loop experiment. Upper plot: reference x and y positions estimated using expected control data $C_x(t)$ and $C_y(t)$; middle plot: actual x and y positions estimated using response surface model control data $C_{1x}(t)$ and $C_{1y}(t)$; and bottom plot: actual x and y positions estimated using neural-controller control data $C_{2x}(t)$ and $C_{2y}(t)$. — **137**

Fig. 5.14 Histogram of time series plot of control data deviations (errors). Upper plot: deviation errors $e_{1x}(t)$ and $e_{1y}(t)$ estimated from response surface model. Bottom plot: deviation errors $e_{2x}(t)$ and $e_{2y}(t)$ estimated from neural-controller. — **138**

Fig. 5.15 Error percentage of control signal predicted by the response surface model and neural-controller in open loop configuration for x- and y-channel. — **139**

Fig. 5.16 Time series plot of beam displacements on the OPD in open and closed loop control configuration for (a) x- and (b) y-position. — **141**

Fig. 5.17 Comparison of Power Spectral Density (PSD) of time series of position data in open and closed loop control configuration for (a) x- and (b) y-channels. — **141**

Fig. 5.18 Alternative (open (red) and closed (blue) loop control) 10 second average values of absolute values of position displacement on (a) x-channel and (b) y-channel for 900 seconds in a 1 minute interval. — **142**

Fig. 5.19 A portion of beam angle of arrival measurement carried out on 11th January 2014 when z_{offset} is constant. — **143**

Fig. 5.20 Laser beam spot centroid motion with and without beam AoA compensation control: (a) beam wandering on 2D-4Q plane with horizontal and vertical scales of real space. (b) Histogram of plane center to beam centroid hypotenuse distance. — **144**

Fig. 5.21 Time series plot of effective scintillation index measured at the receiver station (after propagating 0.5 km) using ensemble average of irradiance of optical signal with beam wandering compensation control turned on (red) and off (blue). — **145**

Fig. 5.22 Normalized impulse response of samples average verses impulse time with and without beam steering. — **146**

Fig. 6.1 Comparison of range of RF, fiber, and FSOC in terms of available data rates against link ranges (Ghassemlooy et al. 2012a). — **151**

Fig. 6.2 Comparison of bandwidth cost for RF, fiber and FSO based communication system (Willebrand & Ghuman 2002). — **152**

Fig. 6.3 Overall layout of constructed free space optical communication data link with the illustration of buildings and playground in between Tower (Tx) and college Information Technology (IT) block (Rx). — **153**

Fig. 6.4 Eye diagram and the distribution of received bits '0' and '1' for BER evaluation. — **158**

Fig. 6.5 Schematic diagram of the free space optical communication experimental setup for 0.5 km data link. On the left is the receiver with beam steering optoelectronic assembly and ob the right is the transmitter. — **160**

Fig. 6.6 Normalized received power against a range of transmittance with and without Beam Wandering Compensation (BWC) control. — **163**

Fig. 6.7 Postprocessed eye diagrams for same modulation scheme with BWC control (a) off and (b) on conditions. (c) Q-factor versus transmittance for OOK-NRZ scheme: theory (solid lines) and experiment (dots). — **164**

Fig. 6.8 BER against a range of transmittance of an FSOC channel. — **165**

Fig. 6.9 Link margin values against the transmittance estimated under different atmospheric turbulence strength conditions with BWC control. — **165**

Fig. 6.10 Illustrations of beam profile when beam wandering (a) mitigated and (b) unmitigated after propagating 0.5 km. The axes are compressed mm scale. — **166**

Fig. 6.11 Geometrical interpretation of beam wavefront (global) profile on the detector surface after propagating 0.5 km horizontal optical path under different turbulence conditions. Beam spot and radial distance are illustrated by red and green colors respectively. — **167**

Fig. 6.12 Experimental Q-factor and theoretical BER estimation against beam centroid displacement on OPD. —— **168**

Fig. 6.13 BER as a function of Q-factor. Theoretical (black) and measured BER with beam steering feedback turned off and on alternately for every 60 s time interval. —— **169**

Fig. 6.14 Experimental Q-factor variations against a range of C_n^2 with beam steering system off and on conditions. —— **170**

Fig. 6.15 Histograms of OOK-NRZ received signal with beam wandering compensation control (a) off in weak and (b) on in strong atmospheric turbulence conditions. —— **171**

Fig. 6.16 BER versus estimated a range of C_n^2 with beam wandering compensation control on and off conditions. —— **172**

Fig. 6.17 Measured BER in 600 runs with alternately beam wandering compensation turned on and off conditions with the measurement interval of 60 s. The horizontal lines indicate the lower and upper boundary of measured BER in both cases. —— **173**

List of symbols and abbreviations

Symbols

U_{b2}^2	Accuracy of the master instrument
O_{pj}	Actual output value of jth neuron
h	Altitude
$\bar{x}$	Arithmetic mean value
L_{tur}	Atmospheric turbulence losses
C_n^2	Atmospheric turbulence strength
$P_r(R)$	Average optical power available at the receiver plane (P_r) at a distance 'R' km
x_{dist}	Beam centroid displacement on x-axis
y_{dist}	Beam centroid displacement on y-axis
γ	Beam centroid Radial Displacement
W	Beam waist
C	Calibration constant
CO_2	Carbon dioxide
C_w	Class wind speed
R^2	Coefficient of determination
U_c	Combined uncertainty
C_x	Control signal for x-axis
C_y	Control signal for y-axis
R	Correlation coefficient
K	Coverage factor
$\phi_h(\xi)$	Dimensionless potential temperature gradient
$\phi_m(\xi)$	Dimensionless wind shear
L	Distance
T_d	Due point temperature
ξ	Eddy dissipation rate
⟨ ⟩	Ensemble average
U_e	Expanded uncertainty
B_j^1	First neuron bias in j^{th} layer
F_1	Footer 1
u	Friction velocity
A_0	Ground level turbulence strength
H_1	Header 1
H_2	Header 2
H	Heat flux
U_{b4}^2	Instrumental error
R	Length of optical link in km
L_m	Link margin
ρ	Mass density
U_{b1}^2	Master instrument uncertainty
Q_{max}	Maximum quality factor
T_{max}	Maximum transmittance
Q	Mean specific humidity
W	Mean vertical velocity
I	Measured irradiance of optical wave

RH_{meas}	Measured relative humidity
T_{meas}	Measured temperature
x_i	Measurement elements
T_b	Minimum pulse duration/bit time
A	Nominal value of C_n^2
σ_{Ex}	Normalized error values for x-position displacement
σ_{Ey}	Normalized error values for y-position displacement
W_{ji}	n^{th} layer weight of neurons i to j
n	Number of samples
b_1	Obukhov buoyancy length scale
U_o	Offset speed
L_{geo}	Optical Geometric Losses
O_2	Oxygen
P	Pasquill stability category
V_{pp}	Peak to peak voltage
V_{Ex}	Position error on x-axis
V_{Ey}	Position error on y-axis
P, Pr	Pressure
C_r	Radiation class
P_R	Received power
V_{Rec}^*	Reference received voltage
B_i	Regression coefficients
σ_1	Relative variance of optical intensity
U_{b3}^2	Resolution of the test instrument
r	Root mean square wind speed
Z_r	Roughness length
σ_R^2	Rytov variance
S_{pd}	Sensitivity of photodiode
R	Solar irradiance
λ	Source wavelength
β_λ	Specific attenuation
C_P	Specific heat
$U(\bar{x})$	Standard uncertainty
V_{Ref}	Sum of energy of four quadrants of OPD
T_{pj}	Target value of jth neuron
T, *Te*	Temperature
ϕ_h	Temperature gradient
T in °C	Temperature in degree Celsius
THz	Tera Hertz
T_{th}	Transmittance threshold
P_T	Transmitted power
O_3	Trioxide
K_h	Turbulence exchange coefficient for heat (or) eddy diffusivity of heat
K_m	Turbulence exchange coefficient for momentum (or) eddy diffusivity of water vapor
V	Visibility
K	Von Karman's constant
H_2O	Water
k	Wave number

Φ_m	Wind shear
Ws	Wind speed
$\mu_a(t)$	Wind speed

Abbreviations

ACK	Acknowledgement
AO	Adaptive Optics
AC Plant	Air Cooler Plant
ASE	Amplified Simulated Emission
A/D	Analog to Digital Converter
ANOVA	ANalysis Of VAriance
AoA	Angle of Arrival
ASIC	Application Specific Integrated Circuit
ARL	Army Research Laboratory
ANN	Artificial Neural Network
ATM	Asynchronous Transfer Mode
A_LOT	Atmospheric Laser Optical Test bed
Atten	Attenuation
ABER	Average Bit Error Rate
BWC	Beam Wandering Compensation
BKB	Bendersky, Kopeika and Blaunstein
BPSK	Binary Phase Shift Keying
BER	Bit Error Rate
CS	Channel Select
CBD	Chesapeake Bay Detachment
CORDIC	Coordinate Rotation Digital Computer
CRC	Cyclic Redundancy Check
DRDY	Data Ready
DRRCs	Data Routing Ring Circuits
DOE	Design of Experiment
DCM	Digital Clock Manager
DSL	Digital Subscriber Loop
D/A	Digital to Analog Converter
DD	Direct Detection
DWT	Discrete Wavelet Transform
ESI	Effective Scintillation Index
EMC	Electro-Magnetic Compatibility
EMI	Electro-Magnetic Interference
ESA	European Space Agency
FIR	Far Infra-Red
FSM	Fast Steering Mirror
FCC	Federal Communication Commission
FTTH	Fiber To The Home
FPGA	Field Programmable Gate Array
FEC	Forward Error Control
FSOC	Free Space Optical Communication

FSOL	Free Space Optical Link
FSO	Free Space Optics
GPS	Global Positioning System
GUI	Graphical User Interface
GMT	Greenwich Mean Time
HL	Hidden layer
HL1	Hidden layer 1
HL2	Hidden layer 2
HDTV	High Definition Television
HV	Hufnagel–Valley
IT	Information Technology
IL	Input Layer
IEEE	Institute of Electrical and Electronics Engineers
IM	Intensity Modulation
ITU	International Telecommunication Union
IP-TV	Internet Protocol Television
LCL	Laser Communication Laboratory
LMS	Least Mean Square
LSB	Least Significant Bit
LED	Light Emitting Diode
LoS	Line of Sight
LTI	Linear Time Invariant
LWC	Liquid Water Content
LAN	Local Area Network
LMDS	Local Multipoint Distribution Service
LT	Local Time
MLCD	Mars Laser Communication Demonstration
MISO	Master In Slave Out
MOSI	Master Out Slave In
MSE	Mean Square Error
MAN	Metropolitan Area Network
MN	MidNight
min/max	Minimum and Maximum
MPAC	Mono-Pulse Arithmetic Circuit
MPA	Mono-Pulse Algorithm
MSB	Most Significant Bit
MAC	Multiply Accumulator
NBIOF	Narrow Band Interference Optical Filter
NABL	National Accreditation Board for Testing and Calibration Laboratories
NASA	National Aeronautics and Space Administration
NPL	National Physical Laboratory
NRL	Naval Research Laboratory
NGN	Next Generation Networks
NRZ	Non-Return to Zero
Nn	Noon
OOK	On-Off-Keying
OFAS	Optical Fiber Amplifiers
OPD	Optoelectronic Position Detector
OE	Output Enable

OL	Output Layer
PISOSR	Parallel In Serial Out Shift Register
PAPR	Peak to Average Power Radio
PAE	Percentage of Absolute Error
PC	Personal Computer
PI	Physik Instrumente
PAT	Pointing, Acquisition and Tracking
PIB	Power in the Bucket
PSD	Power Spectral Density
PLC	Power-Line Communication
PCB	Printed Circuit Board
PRBS	Pseudo-Random Binary Sequence
PRM	Pure Reflection Mirror
RF	Radio Frequency
RoFSO	Radio over Free Space Optics
RAM	Random Access Memory
RH	Relative Humidity
RSM	Response Surface Model
RMS	Root Mean Square
RMSE	Root Mean Square Error
SILE	Semiconductor Laser Intersatellite link Experiment
SCK	Serial Clock
SISOCSR	Serial In Serial Out Circular Shift Register
SPI	Serial Peripheral Interface
SNR	Signal to Noise Ratio
SMF	Single Mode Fiber
SF	Solar Flex
SD	Standard Deviation
SAS	Statistical Analysis System
SIM	Subcarrier Intensity Modulation
SLC	Submarine Laser Communication
SAE	Sum of Absolute Error
SSE	Sum of Squared Errors
TFSLSOC	Terrestrial Free Space Line of Sight Optical Communication
TCSA	Total Cross Section Area
TIA	Trans Impedance Amplifier
TRIG	Trigger
2D	Two Dimensional
TWI	Two Wire Interface
UV	Ultra Violet
UWB	Ultra-Wide Band
UP-NRZ	Uni-Polar Non-Return to Zero
UART	Universal Asynchronous Receiver Transmitter
VHDL	Very High Speed Hardware Description Language
VOIP	Voice Over Internet Protocol
WOC	Wireless Optical Communication

1 Introduction

1.1 Background

Free Space Optics Communication (FSOC) is a Line-of-Sight (LoS) communication where a modulated optical laser beam (visible/infrared) is used to transfer high data rates wirelessly through the atmospheric channel (Forin et al. 2010; Hu et al. 2007). The Romans and ancient Greeks (around 800 BC) used polished metal plates as mirrors to reflect light from one point to another for long range communications (Bell 1980). However, sending information through this transmission method was very limited due to the exchange of predetermined messages only, resulting in low information capacity (Bouchet et al. 2006). In 1792, an optical telegraph based on a chain of semaphores was developed by a French naval navigator called Claude Chappe for communication (Dettmer 2001; Ijaz et al. 2013a). The US military also used sunlight-based power devices to send signals from one mountain top to another in the early 1800s. The blinking of light signals has also been used for many years for ship to ship communication (Ronny 2012). During this development period, Alexander Graham Bell constructed a device called the 'photo-phone' in 1880, which was considered the rebirth of optical wireless communication. Sunlight modulated by voice signals using vibrating mirrors and detected using a selenium-based photo cell was used. In his experiment, telephone-based signals through the atmosphere medium over a range of 200 m were successfully transmitted. The restrictions on this work were the crudity of the devices and the intermittent nature of the atmospheric turbidity (Majumdar & Ricklin 2008).

Experimental exploitation of optical devices for high speed FSOC for long distances requires a monochromatic and narrow strong optical beam at the desired wavelength; such a carrier would have not been possible without the invention of the ruby laser in 1960 by Theodore Maiman, which was considered to be the first successful optical laser (Ijaz et al. 2013b). However, after the invention of semiconductor optical lasers by Robert Hall in 1962, the reliability of operation of the FSOC system had increased sufficiently (Goodwin 1970). Today, semiconductor-injection laser diodes are mostly used for long range optical wireless communication systems (Barry 1984; Corrigan et al. 2009). After the invention of these sources, research in FSOC was continued to enhance the system capacity as well as link range and mainly used in military for secure communications in network-centric operational concepts that promote the use of information as fundamental for gaining superiority on the battlefield (Juarez et al. 2006). FSOC has also been heavily researched for deep space applications by NASA and ESA with programs such as the Mars Laser Communication Demonstration (MLCD) and the Semiconductor-laser Inter-satellite Link Experiment (SILEX) respectively (Fletcher et al. 1991; Peng et al. 2008; Toyoshima et al. 2001). In the past decade, near earth FSOC was successfully demonstrated in space between satellites at data rates of up to 10 Gbps (Hemmati 2006; Jooshesh et al. 2012).

Terrestrial FSOC has now been proven to be a viable complementary technology in addressing contemporary communication challenges, most especially the bandwidth/high data rate requirements of end users at an affordable cost. The fact that FSOC is transparent to traffic type and data protocols makes its integration into the existing access network far more rapid, reliable and profitable in comparison to traditional fiber communications (Ghosh et al. 2005). Despite these advantages, FSOC performance is degraded by the substantial optical signal losses due to atmospheric particles absorbing and scattering the propagating optical and infrared waves, since their wavelengths are very close to the wavelengths of these frequencies (Soibel et al. 2009). However, in clear weather conditions, theoretical and experimental studies have shown that scintillation can severely degrade the reliability and connectivity of FSOC links (Gappmair & Flohberger 2009; Nistazakis et al. 2009). Nevertheless, the atmospheric channel effect poses a great challenge to achieve link availability and reliability according to the IEEE and ITU link availability standards of 99.999 % (five nines) for the last mile access communication network. Therefore, these channel effects at the installation spot still need to be understood and circumvented in order to increase the link range and link availability in terrestrial FSOC systems (Zabidi et al. 2010; Wu & Kavehrad 2007; Wakamori et al. 2007; Fischer et al. 2004).

1.2 Research motivation

In the past decade, the world has witnessed a spectacular growth in the traffic carried by telecommunication networks. As the number of users using applications requiring a large bandwidth is increasingly growing, the bandwidth limits of current wireless systems in radio frequency based technologies are being stretched (Mahdy & Deogun 2004; Ijaz et al. 2013b). Recently, FSOC systems with a huge unlicensed modulation bandwidth capability have attracted a great deal of interest from a number of sources including academia, industry, telecommunication and standardization bodies. This huge bandwidth represents high potential in terms of capacity and flexibility thus making FSOC technology a particularly attractive candidate for multi-gigabit wireless applications including audio, video streaming, multi-gigabit file transferring and internet revolution for last mile access network, mobile telephony backhaul (3G), satellite communication offering better quality and user experience, and in areas to complement radio frequency RF-based services (Bouchet et al. 2008; Cvijetic et al. 2009; Kaushik et al. 2012).

FSOC has become increased popular during the last few years due to its increased power efficiency for reasonably longer distances. Moreover, it provides data rates that can cater for our future broadband telecommunication requirements besides resolving the last mile access bottleneck. FSOC technology has attractive characteristics like dense spatial reuse due to light beam directionality (Pederson & Solgaard 2000), low power usage per bit, and license-free band of operation etc. (Majumdar & Ricklin 2008). For more than a decade an increasing demand for high bandwidth transmission capabilities can be noted, which should be flexible and allow quick and easy installation at low costs. Among other technologies, this has attracted much attention for wireless optical point-to-point transmission (Sprangle et al. 2009). This trend is supported by the development of several new optoelectronic components and the increase in data rates in wide area fiber optic networks, which outperform any other transmission technology today in typical ambient conditions (Hemmati 2009).

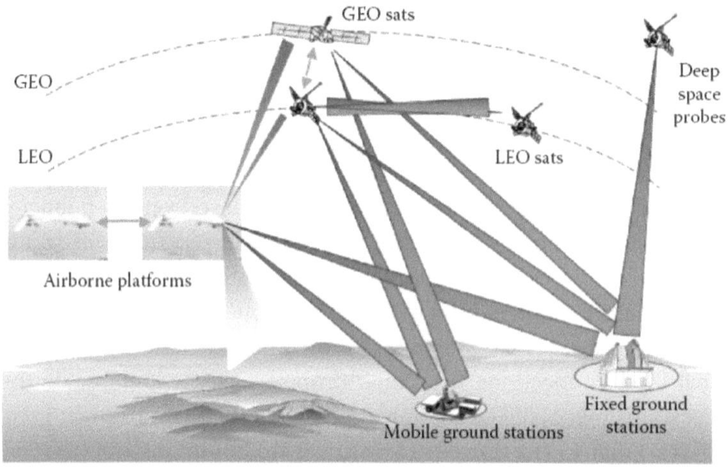

Fig. 1.1: A typical concept of an FSOC network.

An FSOC network for ground-to-ground, ground-to-space, space-to-space and space-to-ground data links is shown in Fig. 1.1. These links may include, for example, a metropolitan network between office buildings, a long-distance wide network between ground stations and satellite platform, or an inter-satellite link (Bloom et al. 2003). Another area of interest is the broadband wireless solution for closing the so-called last mile connectivity gap throughout metropolitan networks. Different network architectures and the usage of optical repeaters, point-to-point, and point-to-multipoint solutions are also possible in FSOC with the presently available optoelectronic components (Dat et al. 2011; liu Yu-Tai et al. 2005).

Recent rapid progress in information and communication technologies has exceeded our expectations for meeting the requirements of a multimedia society in the 21st century. FSOC is considered to be one of the key technologies for realizing very-high-speed full duplex multi-gigabits-per-second (multi-Gb/s) large-capacity throughput aerospace communications for voice, video and data information (Ciaramella et al. 2009). Using lasers as signal carriers, FSOC can provide an LoS, wireless high-bandwidth communication link between remote sites. Rapidly growing use of the Internet and multimedia services (Grant et al. 2006a) has created congestion in the telecommunications networks and placed many new requirements on carriers. Laser transmitters offer an intermediate low-risk means to introduce desired network functionalities with extremely high bandwidth (Awan et al. 2009a). The wireless aspect of FSOC can be a crucial advantage, particularly in Local Area Networks (LANs) and Metropolitan Area Networks (MANs) in cities where the laying of fiber optic cables is expensive/difficult.

FSOC offers substantial advantages over conventional RF wireless communications technology, including higher data rates, low probability of intercept, low power requirements, and much smaller packaging (Tunick 2007a). FSOC systems have proven to be a viable alternative to optical fiber based systems in several applications, as the technology comes closer and closer to providing the 99.999 % service that many corporations require of their data networks. During the last thirty years, great advances have been made in electro-optics and optoelectronics and incorporated into today's FSOC systems, mostly for defense applications (Liu et al. 2010). Furthermore, given the fact that the optical spectrum is unlicensed with frequencies of the order of hundreds of terahertz, FSOC can be installed license-free worldwide. Most FSOC systems use simple ON-OFF keying as a modulation format, the same standard modulation technique that is used in digital fiber optics systems. This simple modulation scheme enables FSOC systems to provide bandwidth and protocol transparent physical layer connections. The other proposed methods to tackle the "last mile" bottleneck are Power-Line Communication (PLC), Digital Subscriber Loop (DSL) or cable modems, Fiber To The Home (FTTH), Local Multipoint Distribution Service (LMDS), and Ultra-Wide Band (UWB) technologies. The limitations and demerits of these methods are (i) more expensive, (ii) easily damageable, (iii) time consuming, (iv) complicated to maintain and reconfigure, (v) dependency on the common radio link, (vi) attenuation and outage during rainfall, (vii) carrier frequency license band issues, and (viii) frequency interference with other system etc. (Ijaz et al. 2013b).

FSOC, a fiberless, laser driven technology, offers similar capacity to that of optical fiber based communication with significant reductions in cost and time. The integration of FSOC into the access network can be done relatively cheaply and quickly as it is transparent to the traffic type and protocols. However, the channel in FSOC poses a great challenge and the performance of an FSOC system is subject to abrupt changes in atmosphere. Therefore, it is desirable to experimentally characterize and analyze the system's performance under the different atmospheric conditions. A number of authors have studied the effect of atmospheric turbulence, however, most of the studies are theoretical, analytical, simulation or indoor laboratory based setups and very little work has been reported with real outdoor experiments (Li & Uysal 2003). This is because in practice, it is very challenging to measure the effect of atmosphere turbulence under diverse conditions (Mndewa et al. 2008). This is mainly due to the long waiting time to observe and experience reoccurrence of different atmospheric events

(Ijaz et al. 2013a). However, when the link length exceeds several hundred meters irradiance fluctuations of the received optical signal due to the turbulence poses a severe problem (Chand 2000).

The turbulence induced by the random fluctuation of temperature and pressure results in random variation of the atmospheric refractive index. The variations in the refractive index along the optical path cause random fluctuations to the received optical irradiance, which can lead to severe system performance degradation (Bazil Raj et al. 2010). Very little work on the Bit Error Rate (BER) performance of an FSOC link in a real outdoor link over a longer period has been reported. Therefore, this research work characterizes the atmospheric turbulent channel in different weather conditions together with the profile of the data carrying optical beam (Qingfang et al. 2011). The dependence on atmospheric channel effects and mitigation techniques is considered to be an important parameter in order to achieve maximum transmission span (Bazil Raj et al. 2011). Despite the advantages of FSO links, operating them at THz and FIR wavelength bands would require high cost components that are not readily available at the moment. Therefore, almost all commercially available FSO systems operate in the wavelength range of 0.60–1.55 μm.

1.3 Characteristics of FSOC

The main characteristic properties of FSOC are as follows:

1.3.1 Directionality of the light beam

The light beams used for FSOC are much narrower and typically have an angular width of 1 mrad, as opposed to omni-directional RF, which occupies 360° in a plane. Because of this, there are generally no interference issues in FSOC and there are no or very little medium access issues (Ali 2013a). Directionality also helps in localization, because it is very easy to get orientation information from the neighbor, unlike wireless RF networks.

1.3.2 Form factors, i.e. size and power per bit

The size of the equipment used for short range FSOC can be small i.e. a few centimeters. Semiconductor lasers and LEDs used for FSOC are of very little power (a few milliwatts) which makes FSOC suitable for power limited ad-hoc, sensor network and multi-hop network scenarios to improve the link quality and reduce the Peak to Average Power Ratio (PAPR) (Majumdar & Ricklin 2008).

1.3.3 Ability to be operated license-free worldwide and quick installation

Optical wavelengths are license free and FSOC deployment does not require any permission as long as they are eye safe. FSOC systems can be deployed in an ad-hoc manner and typically can be installed in a single day (Barrios & Dios 2012). Also, the system can be made to operate behind transparent windows, avoiding expensive rooftop rights.

1.3.4 Wavelength selection criteria

The choice of transmitting laser wavelength will depend on the atmospheric propagation characteristics, optical background noise, and the technologies developed for lasers, detectors, and spectral filters. For a long atmospheric channel, the wavelength will generally need to be restricted to spectral regions of very low atmospheric absorption. Another consideration is intensity fluctuation (scintillation) due to turbulence. The strength of intensity fluctuations decreases as $\lambda^{-7/6}$: thus, scintillation and hence BER can be decreased by using a longer wavelength (Hemmati 2009).

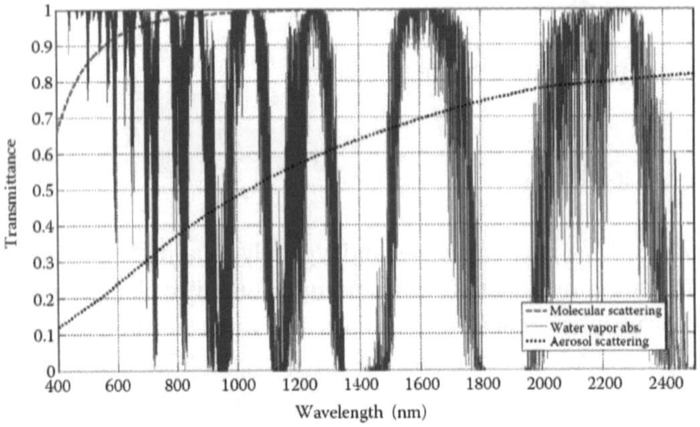

Fig. 1.2: Contribution of Rayleigh scattering, Mie scattering and water vapor to atmospheric transmittance.

Figure 1.2 shows the variation in atmospheric transmittance resulting from molecular scattering, water vapor absorption and aerosol scattering by various components (O_2, H_2O, CO_2, O_3) according to wavelength in the 400 to 2600 nm range. Notice that the band structure is mainly due to water vapor. Effects of molecular scattering are not noticeable over 1000 nm while aerosol scattering is noticeable all over the spectrum (Hemmati 2009).

1.3.5 Challenges and limitations

In order to ensure the quality and reliability of data link access through FSOC technology, an accurate characterization of data transmission for different environmental conditions is of utmost importance. In case of strong atmospheric turbulence, the optical beam undergoes serious challenges in terrestrial FSOC systems and the data rate it could operate is degraded. The major challenges are (i) real-time study of atmospheric turbulence strength at the location where the system is installed, (ii) beam stabilization on the detector plane, and (iii) wavefront distortion equalization (Perez-Arancibia et al. 2012; Plett et al. 1999).

1.4 Applications and advantages

Some of the typical scenarios where FSOC could be employed are given as follows:
- LAN to LAN connections on campuses/in city at fast-ethernet or gigabit-ethernet speeds
- Speedy service delivery of high bandwidth access to fiber networks
- Temporary network installation (for special events or other purposes)
- Re-establishing high-speed connection quickly (disaster recovery and emergency response).
- Communications between ground and spacecraft, or between spacecrafts, including elements of a satellite constellation
- Interstellar communication
- Ship-to-ship communications with high data rates providing complete security
- Military/defense communications

Some typical advantages possible with FSOC are:
- High data rates
- High transmission security
- No Federal Communications Commission (FCC) license or frequency allocation
- Light weight, small volume, and lower power consumption, providing a potential edge over RF communication
- Portability and quick deployment
- Increased security due to the laser's narrow beam – ideal for the wireless transfer of financial, legal, military/defense and other sensitive (highly secured) information

1.5 Research objectives

The prime aim of this research work is as follows:
1. To develop a model that relates the received optical signal fluctuation with the atmospheric turbulence changes based on the experimental data for modeling and simulation of atmospheric characteristics.
2. To develop and demonstrate mono-pulse, Artificial Neural Network (ANN) and parallel processor based steering techniques in Terrestrial Free Space Line of Sight Optical Communication (TFSLSOC) on a test bed with a simplex communication link for a range of 500 m.

1.6 Original contributions – newness and achievements

As a direct result of this research, the following original contributions have been made:
1. The experimental setup, wireless optical transmitter and receiver system with desired optoelectronic devices has been erected for 0.5 km link range at an altitude of 15.25 m through which the entire research work is carried out.
2. A new low cost dynamic measurement system is developed with high accuracy of correlation coefficients of 99.92 %, 99.63 %, 99.73 % and 99.88 % for wind speed, temperature, relative humidity and pressure respectively, used to continuously acquire the meteorological data which are the only input to develop the model and characterize the turbulence channel.
3. Separate models are developed for predicting the atmospheric attenuation and turbulence strength as a function of local meteorological data acquired in various outdoor environmental conditions (since optical wave propagation in the atmosphere is seasonally dependent) and prediction accuracy of 0.041 dB/km for atmospheric attenuation and $0.000631 \cdot 10^{-9}\,m^{-2/3}$ for turbulence strength are achieved and validated against the selected existing models.
4. The optical beam fluctuation due to atmospheric turbulence and dynamic disturbance is mitigated using two different controllers (developed based on Response Surface Model (RSM) and ANN) implemented in the MATLAB environment at the receiver station. The performance of the developed controllers is intensively tested in real time and an outstanding behavior from the neural-controller is achieved.
5. A pipelined-parallel digital architecture is developed in Field Programmable Gate Array (FPGA) according to the proposed software neural-controller structure and the timing performance in terms of correction speed and accuracy is significantly improved.
6. The quantitative analysis of FSOC data transmission quality and reliability metrics are measured with and without a beam Pointing Acquisition and Tracking (PAT)

system through which the maximum possible data rate at which the system could operate is characterized.

1.7 Thesis organization

The thesis has been organized in seven chapters as follows:

Chapter One – Introduction: A complete foreword covering FSOC technology, research motivation, characteristics, typical applications and advantages is presented together with its distinctive features. It also consists of research objectives as well as its original contributions.

Chapter Two – Real-time measurement of meteorological parameters for estimating low-altitude atmospheric turbulence strength (C_n^2): A complete overview of meteorological parameter influence on the propagating laser beam and the state-of-art literature review are presented. The detailed description of the configuration of the field test experimental setup is introduced and the measurement technique is outlined. The specialized weather sensors interfacing technique, simulation results and data conversion methods are described. The performance calibration of the proposed measuring instrument is detailed along with uncertainty computation results. A more common method of atmospheric turbulence strength estimation (Tunick 2005; Oh et al. 2004a; Doss-Hammel et al. 2004; Majmudar et al. 2008) is reviewed and the final formula is interpreted. The measurement and estimation results acquired in different seasons for the diurnal period are discussed along with the associated meteorological parameter variation data.

Chapter Three – Modeling and numerical evaluation of ground-level atmospheric attenuation and refractive index structure function (C_n^2) using measured local meteorological data for FSO applications: The main objective of this chapter is developing separate models to make more accurate estimations of atmospheric attenuation (Aatt) and turbulence strength (C_n^2). Backgrounds and related works are reviewed and the results are presented. Experimental setups and instrumentation assembly on the vibration damped optical breadboard are described. Various existing models developed based on the theoretical, analytical, empirical and experimental approach to estimate the Aatt and C_n^2 are reviewed. The development of new models based on the analysis of variance (ANOVA) method as a function of local meteorological data, fitting for different local outdoor environmental conditions, are described. The comprehensive analyses carried out on the measurement and estimation values associated with different weather seasons and the results in terms of estimation accuracy, Root Mean Square Error (RMSE), is discussed. It also highlights the estimation ability of the new proposed models.

Chapter Four – Mitigation of beam wandering due to atmospheric turbulence and prediction of control quality using intelligent decision-making tools: This chapter outlines various atmospheric effects on the propagating optical beam and the method of stabilizing the center of the beam at the detector plane. The state-of-art and literature review results are presented. The optoelectronic assembly and setup of FSO link (unmodulated beam) established are described and the inherent linear/nonlinear behavior of individual parts is highlighted. The design approach of the Response Surface Model (RSM) and the neural controller model are described. The experimental confirmatory test results associated with the control ability of these controllers in terms of time and accuracy are discussed and the improved performance of FSOC link reliability is analyzed.

Chapter Five – A direct and neural controller performance study with beam wandering mitigation control in free space optical link: This chapter introduces the necessity for hardware realization of the proposed controllers. The background and related works are presented. The experimental plant configuration layout with aerial view and data handling and manipulation hardware design flow is explained. The design and implementation of the proposed controllers in FPGA is described. The hardware controller performance in beam centroid stabilization in open and closed loop configuration using the developed test bed is intensively tested and associated results are analyzed.

Chapter Six – Quality metrics and reliability analysis of ground-to-ground free space laser communication in different weather conditions together with beam steering system: In this chapter, the parameters corresponding to the analysis of the quality and reliability of FSOC data transmission are introduced. The overviews of literature survey results are presented. Further, this chapter deals with the quantitative measurement of data transmission quality during different atmospheric attenuation and turbulence strength that are estimated using proposed models; through which the performance improvement of the FSOC system by incorporating the beam steering controller is studied. These estimation and measurement results, corresponding to different weather data, are explained. The experimental results and characterization of maximum data rate that the system could operate in various outdoor weather conditions with and without PAT are presented in detail.

Chapter Seven – Conclusions and future work: Summary of newness and findings are presented in this chapter. The conclusions as well as future work are outlined.

The overall contribution of the thesis is schematically illustrated using a research roadmap as depicted in Fig. 1.3.

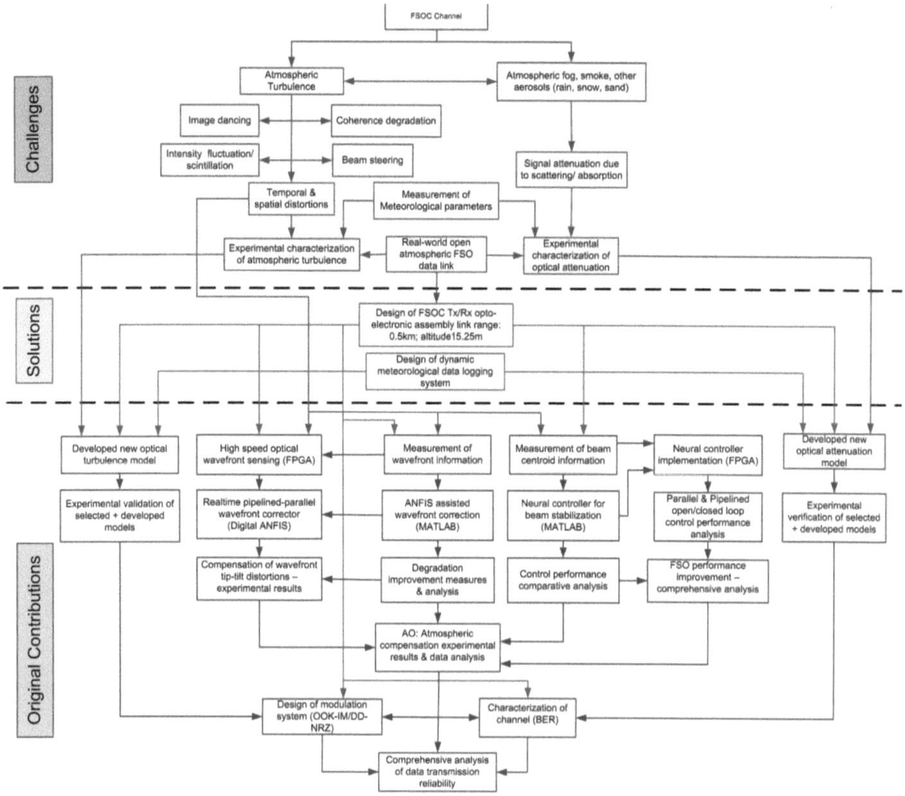

Fig. 1.3: Outline of the research roadmap and original contribution.

2 Real-time measurement of meteorological parameters for estimating low altitude atmospheric turbulence strength (C_n^2)

The major factor that limits the performance of Free Space Optical Communication (FSOC) is atmospheric turbulence which fluctuates over time in accordance with the variations in local meteorological parameters. Estimating the atmospheric turbulence strength (C_n^2) from measurement data is significant for finding the data rate the system is capable of operating at under different outdoor local environmental conditions. Hence, a low cost customized system for continuously measuring local meteorological data is developed and presented in this chapter. A field test scintillometer setup is established for a link range of 0.5 km at an altitude of 15.25 m. Specialized sensors are interfaced to the digital architectures to acquire the real-time data corresponding to atmospheric changes. The accuracy and performance of the measurement system are tested against standard instruments and the maximum correlation coefficients of 99.92 %, 99.63 %, 99.73 %, and 99.88 % are achieved for wind speed, temperature, relative humidity and pressure respectively. Atmospheric turbulence strength is estimated for the diurnal period using measured meteorological data. The validations of the estimated results with the scintillometer measurement are also analyzed. The weather profile and corresponding C_n^2 variations at our test field for different seasons over a one year period are presented and the results are analyzed.

2.1 Introduction

Free Space Optical Communication (FSOC) is an alternative emerging technology to meet future requirements and Next Generation Networks (NGN) demands. FSOC is identified as an alternative to complement microwave (mm wave) and Radio Frequency (RF) links within the access network for backhaul traffic (Awan et al. 2009b). The performance of an FSOC system is examined in two different categories: (i) internal parameters (optical power, wavelength, transmission bandwidth, divergence angle, optical loss, receiver sensitivity etc.) and (ii) external parameters (visibility, atmospheric attenuation, turbulence strength, scintillation, deployment distance, pointing loss etc.). All these parameters are not independent but are linked together in specifying overall system performance. The FSOC links are also influenced by atmospheric temperature that varies both in spatial and temporal domains. The variations of temperature in the FSOC channel are functions of atmospheric and geographical parameter variations. This effect is commonly known as optical turbulences or scintillation effects (Font et al. 2006; Sitterle et al. 1988). Atmospheric scintillations can be defined as the changing of light intensities in time and space at the plane of a receiver that detects a signal from a transmitter located at a distance (Bao et al. 2012). The received signal at the detector fluctuates as a result of the thermally induced changes in the index of refraction of the air along the transmit path (Awan et al. 2009c). The time scale of these fluctuations is in the order of milliseconds, approximately equal to the time a volume of air takes to move across the path of beam size; and therefore is approximately related to wind speed (Bloom et al. 2003). Overall, scintillation causes rapid fluctuations of received power and in the worst case results in high BER (Bloom et al. 2003). Maintaining a clear LoS between transmitter and receiver terminals is the biggest challenge in establishing FSOC data links in the troposphere (Awan et al. 2009b). As the near ground FSOC system is deeply affected by atmospheric turbulence (Santiago et al. 2005), it is very important to analyze the channel behavior in different sessions (Liu et al. 2010). The continuous measurement of atmospheric turbulence strength and its effects on the laser beam propagation in different environmental conditions over a long period becomes significant (Gappmair 2011) for analyzing the quality and reliability of an FSOC system and the maximum bit rate the system could operate. Hence a low cost, precise and customized atmospheric parameter measurement system is developed and presented in this chapter. Atmospheric forecasting is conducted by collecting the real-time wind speed, temperature, relative humidity and pressure data using Cup Anemometer, Temperature and Relative Humidity Sensor (SHT11) and Absolute Pressure Sensor (SCP1000-D01) respectively. The sensor units are interfaced with the digital architecture developed in the Xilinx-Virtex-5 LX50T FPGA platform. A proper synchronization is maintained between the architecture and the sensors by the clock manager for precise on-time data conversion and acquisition process.

The rest of the chapter is organized as follows: Section 2.2 presents the background and related works, Section 2.3 describes the experimental test bed and measurement protocol, Section 2.4 explains the pipelined digital architecture developed for sensor interfacing and data acquisition process, Section 2.5 briefs explains the RS-232 Universal Asynchronous Receiver Transmitter (UART) digital circuit and communication frame format, Section 2.6 describes the proposed measurement system calibration results along with the relevant uncertainty in detail, Section 2.7 presents the overview of atmospheric turbulence strength estimation, Section 2.8 discusses the experimental results and data analysis, Section 2.9 presents the advantages of the proposed measurement system and Section 2.10 draws the summary.

2.2 Background and related works

In the last few years, a lot of in situ field measurements related to environmental parameter monitoring and C_n^2 measuring have been carried out and can be found in the literature. An overview of closely related works is given in this section.

The theory of atmospheric turbulence related to the index of refraction model, the Rytov method, intensity fluctuation, and aperture averaging are described by Peng Liu et al. (2010) and the seasonal changes of C_n^2 are presented. It is concluded that beam steering is significant to mitigate the Angle of Arrival (AoA) error.

Steve Doss-Hammel et al. (2004) describe the field test experimental setup and test protocol used for estimating C_n^2. The ascertainment of C_n^2 using the PAMELA model is evaluated for an optical horizontal path over land and water. The results are presented and analyzed.

Yahaya and Frangi (2004) describe some dynamic characteristics of the optical and cup anemometers in terms of spectral intensity, frequency, wave number and power spectra. The experimental approach to measure the natural wind turbulence by both anemometers is discussed. The cup anemometer for long-term wind velocity measurement is suggested.

Fred J. Taylor et al. (1977) propose a digital automated wind measurement system. LED and phototransistors (optoelectronic conversion system) are used to count the driver shaft revolution. Laboratory, field experimental results and calibration errors are presented.

Pelegri-Sebastia et al. (2012) propose a Relative Humidity (RH) measurement method based on microcontroller and capacitive type RH sensors. An artificial neural network is used to linearize the sensor's response and to reduce the external hardware. A flowchart for capacitance measurement is given and explained.

Vladutescu et al. (2012) describe a community multiscale air quality model to provide air quality predictions that can be used for forecasts, and for a better understanding of the interplay of meteorology, atmospheric emissions and chemistry. Mie scattering and the effects of relative humidity are used to get a vertical profile of aerosol distribution. The results are presented and analyzed.

Tunick (2007b) describes an experiment developed for a 2.33 km near horizontal optical path and presented the results calculated using Rytov variance. The spectral analyses for measured laser signal intensity data are shown. The comparison results of scintillometer and calculated C_n^2 data are explained.

Hunt (1999) presents empirical, statistical and combined modeling techniques for environmental forecasting and atmospheric turbulence with a forecasts construction sketch and calculation grid.

Fernando Lopez Pena and Richard J. Duro (2003) present an automatic calibrator developed for fast and accurate calibration of anemometers. An Artificial Neural Network (ANN) aided virtual environment with many sensors is created to increase the accuracy of calibration even by inexperienced users. The performance improvement is highlighted by means of achieved uncertainty.

Oh et al. (2004b) present the environmental changes and optical turbulence estimation results calculated using the PAMELA model in different sessions. The humidity effects are analyzed. It is concluded that humidity and turbulence are inversely proportional.

Arnold Tunick et al. (2005) present an overview of selected optical turbulence (scintillometer data) and meteorological data collected at the Army Research Laboratory (ARL) and Atmospheric Laser Optical Test bed (A_LOT) facility on different durations over 2.3 km elevated optical path.

M. Kusnerova et al. (2013) discuss various methods for simplifying the method of evaluation of uncertainties in measurement results.

The C_n^2 is modeled as a function of altitude (h) in all the models except the Hilbert–Huang decomposition, the bulk method (Frederickson et al. 1998), the Hufnagel–Valley model (Majumdar et al. 2008) and the PAMELA model (Oh et al. 2004).

The Hufnagel–Valley model estimates the C_n^2 as a function of wind speed (Ws), ground level turbulence strength (A_0) and altitude (h). Most of the other models are based on optical turbulence similarity theory and predict similar results; however, the local meteorological and geographical parameters are not included but are the input to the PAMELA model. The C_n^2 could be estimated as a function of local meteorological and geographical data using the PAMELA model (Doss Hammel et al. 2004).

Mazin Ali A. Ali (2013b) explain the Rytov variance for plane and spherical waves. Results of atmospheric turbulence effects on wavelength transmission (1550 nm, 850 nm, 633 nm, 532 nm) in free space are given and analyzed. Scintillation attenuations and log SNR are computed for different propagation distances and the results are discussed.

2.3 Field test experimental setup and measurement protocol

A simplex Free Space Optical Link is established for the range of 0.5 km at an altitude of 15.25 m in a Laser Communication Laboratory (LCL). A scintillometer experiment test setup is constructed with necessary optoelectronics components to measure C_n^2 field data (as a direct measurement) near the horizontal optical path as shown in Fig. 2.1. The transmitting and receiving setups are mounted on vibration damped optical breadboards.

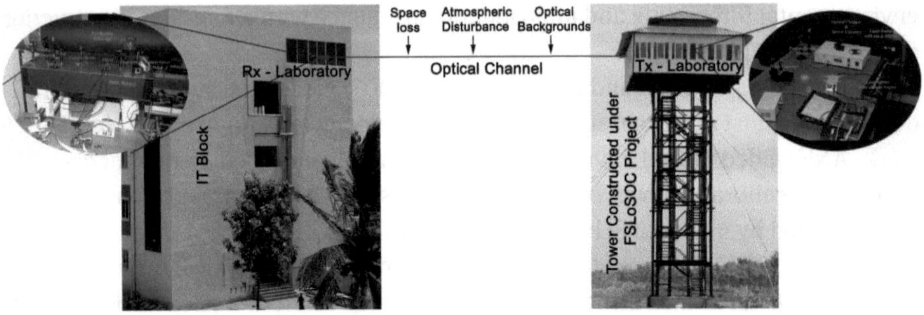

Fig. 2.1: Optical propagation path for the local C_n^2 field data measurement and acquisition with direct transmission test equipment (scintillometer setups). The receiver system is seen on the building at the left and the transmitter system is seen on the tower at the right.

Tab. 2.1: Parameters of the optical link.

Parameter		**Value**
Transmitter		
Laser diode	Peak wavelength	850 nm
	Maximum optical power	10 mw
	Beam size at aperture	3 mm
	Beam divergence	3 mrad
	Laser beam propagation model	Plane
Optical Lens	Diameter	3.9 mm
Channel	Range	0.5 km
	Altitude	15.25 m
	Surface roughness length	0.03 m
	Wind speed	0–38 m s^{-1}
	Temperature	22–58° C
	RH	0–100 %
	Pressure	99–101.9 kPa
Receiver		
Telescope	Aperture	330.9 mm
	Type	Newtonian
Filter	Type	NBOIF
	CWL	850 nm
Optical collimator	Collimation ratio	9 : 3
Photodetector	Active area	1 mm^2
	Half angle field of view	±75°
	Spectral sensitivity	0.59 A/W
	Rise and fall time	5 ns
Data processing	FPGA	Virtex5
Data logging	PC	Quad processor

The main optoelectronic devices and their parameters are given in Tab. 2.1. The Rytov variance is a measure of the strength of scintillation. The ray propagation through a turbulent atmospheric medium will experience irradiating fluctuations called scintillation (Grant et al. 2006b). The relation between turbulence strength (C_n^2) and the relative variance of optical intensity σ_I^2 was set by Rytov as (Tunick 2007; Ali 2013b)

$$\sigma_1^2 = KC_n^2 k^{\frac{7}{6}} L^{\frac{11}{6}} , \tag{2.1}$$

where C_n^2 is the turbulence strength parameter, k represents the wave number ($k = 2\pi/\lambda$), L is the distance between the transmitter and the receiver of the optical wireless link and K is a constant (K = 1.23 for plane wave approximation and 0.5 for spherical wave approximation). The signal intensity (scintillation) data at the receiver is continuously recorded and the scintillation index σ_I^2 is obtained by

$$\sigma_1^2 = \frac{\langle I^2\rangle - \langle I\rangle^2}{\langle I\rangle^2}\,, \tag{2.2}$$

where I is the measured irradiance of the optical wave and the angle brackets $\langle\cdots\rangle$ denote an ensemble average or equivalently a long-time average. From Eqs. (2.1) and (2.2), the C_n^2 can be calculated for the 850 nm optical source and 0.5 km link range as

$$C_n^2 = \frac{\sigma_1^2}{1.23k^{\frac{7}{6}}L^{\frac{11}{6}}}\,. \tag{2.3}$$

σ_I^2 is computed from the irradiance of the optical wave observed by a point detector after propagating a distance L (Scintec Corporation 2012). Classical studies on optical wave propagation have been classified in two major categories, either the weak or strong fluctuations theory. It is customary to discriminate both cases for a given propagation problem by determining the value of Rytov variance. The weak fluctuations regime occurs when $\sigma_R^2 < 1$ and the strong fluctuations regime associates with $\sigma_R^2 > 1$, while there is a saturation regime when $\sigma_R^2 \to \infty$.

The weather sensors are mounted on the PCB and placed inside a sensor house in order to prevent them from the direct influence of external heat and rain impact as shown in Fig. 2.2. A rooftop with membrane is built to improve the response time and accuracy of the wind speed (Ws), temperature (T), relative humidity (RH) and pressure (Pr) sensors. The side apertures of the Ws sensor section are open, so that the sensor is influenced by the wind irrespective of the direction. This weather station is positioned near the scintillometer receiver setup. The meteorological parameters are measured every second (at the rate of 1 Hz) and one minute average data are recorded in the PC for several diurnal periods in different seasons.

2.4 Sensor interfacing architectures and data acquisition protocols

The weather sensors are connected to an FPGA in which the global pipelined-parallel interfacing architecture is developed. A simple pulse counting, Two Wire Interface (TWI) and Serial Peripheral Interface (SPI) protocol are used for proper synchronization and data acquisition from sensors. Very High Speed Hardware Description Language (VHDL) is used for programming to develop the digital architectures inside the FPGA. The hardware interfacing circuit operations and data acquisitions are described in this section.

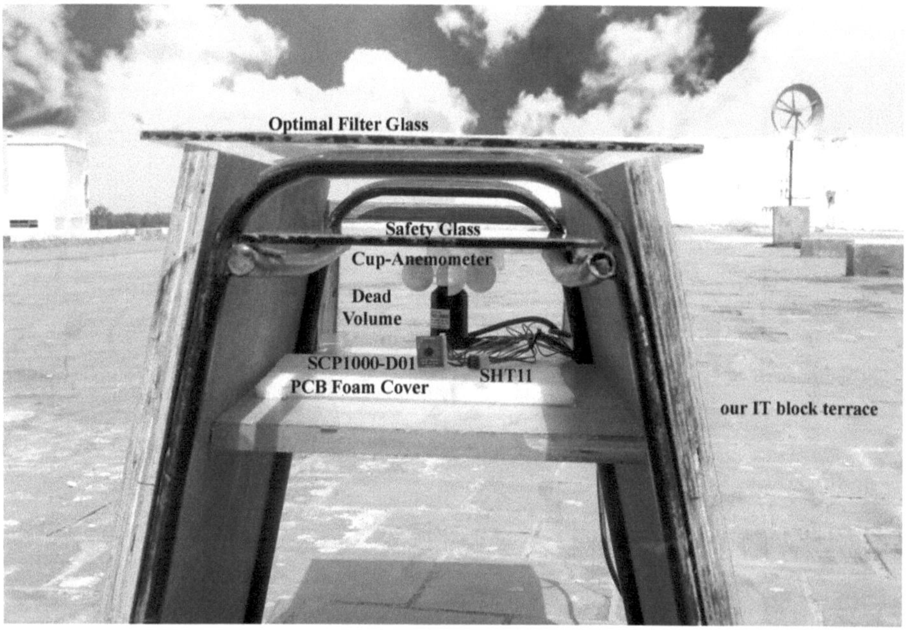

Fig. 2.2: Photograph of sensor compartment – sensors are protected from the direct influence of environmental changes and kept in a waterproof housing.

2.4.1 Wind speed measurement – cup anemometer

The cup anemometer assembly for wind speed measurement consists of hemispherical cups which rotate according to wind speed, a drive shaft, magnetic bars and a single pole reed switch. The angular movement of the magnet is directly proportional to the rotational speed of the cups i.e. wind speed. The reed switch is closed when any one of the magnetic poles comes closer to its center part and opens when the pole moves away as shown in Fig. 2.3. The number of contacts (pulses) per second is the measurement of wind speed (Fasinmirin et al. 2011). There is a minimum wind speed which will set the cup in motion depending on the friction in the bearings of the wheel and the design parameters of the instrument (Coquilla 2010). In a steady wind at the LCL, the cup performs well from almost 0.27 $m\,s^{-1}$ to 60 $m\,s^{-1}$. This is because the cup wheel, having inertia, accelerates more rapidly with an increasing wind speed than it decelerates with decreasing wind speed (Pena & Duro 2003).

The wind speed $\mu_a(t)$ is related to the angular velocity of the anemometer by (Yahaya & Frangi 2004)

$$\mu_a(t) = Cs(t) + U_0 , \qquad (2.4)$$

where C is the calibration constant (gain = 0.6201 m), s(t) is the angular velocity of the device in Hz and U_0 is offset speed. A JK Flip-Flop is used to hold the pulse from the anemometer and to reset it after reading the pulse. The anemometer interfacing

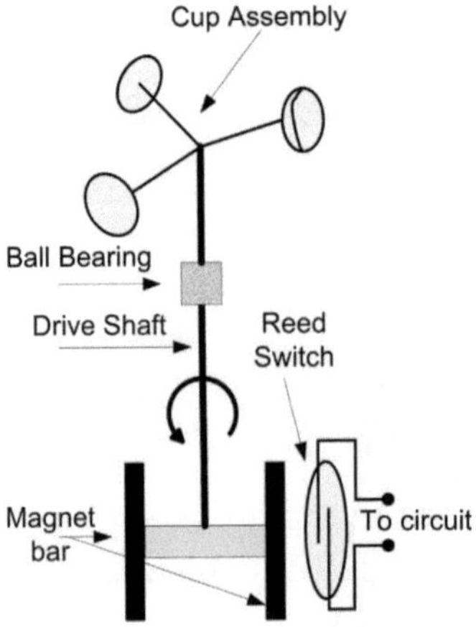

Fig. 2.3: Components assembly and design structure of cup anemometer – kept in a waterproof cabin.

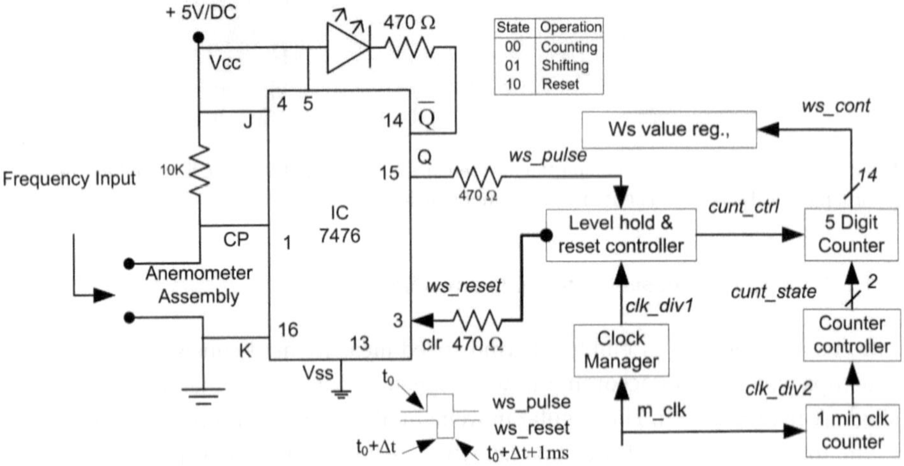

Fig. 2.4: Circuit schematic and architecture of cup anemometer configuration for wind speed measurement.

circuit and digital architecture is shown in Fig. 2.4. Every poling of the anemometer is actually a clock pulse to the JK FF and the pin 'Q' goes high for every poling. The "level hold & reset controller" continuously monitors the level changes at pin 'Q' and increases the value of a 5-digit counter at every rising edge. After one millisecond, it resets the JK FF by sending a low signal to the 'clr' pin. The LED glows every time the anemometer sends a signal to the FPGA.

A one second clock counter is designed to trigger the counter controller unit once in every second, which keeps the counter in counting mode, shifting mode or reset mode. The clock manager generates the control and trigger signal at 5 kHz rate in order to synchronize the measurement. The counter value $(000000010000)_2$ for a given second, is equal to 16 in radix 10. The Reed switch of the anemometer gets triggered four times per cycle, hence the angular velocity s(t) of the anemometer is given by counter value/4. The offset wind speed of the anemometer and the calibration constant are 0.27 m s^{-1} and 0.6201 m respectively. The pseudocode of measurement algorithm and simulation timing response are shown in Figs. 2.5 and 2.6 respectively.

```
Declaration:        ws_pulse, m_clk:=in: =0;
ws_reset:=out:=0;
ws_cunt, ws_value:=wire=[13:0]:=0
clk_div1, clk_div2, cunt_ctrl:=wire:=0;
cunt_state:=wire:=0;
i,j,k,l,n :=variable :=integer:=1;
Process (1&2):While (1)
clk_div1<= not clk_div1, i:=1 when i:=l else i:=i+1
if ws_pulse:=1, cunt_ctrl<=1 else cunt_ctrl<=0;
wait(), ws_reset<=0 wait () ws_reset<=1;
Process(3): If cunt_ctrl=1, if cunt_stat:=0,
ws_cunt<=ws_cunt+1;
elsif cunt_state:=1, ws_value<= ws_cunt;
elsif cunt_state:=2, ws_cunt<=0;
else ws_cunt<=ws_cunt+1; endif; endif
Process(4&5): Clk_div2<=not clk_div2, k=1 when k:=m;
else k:=k+1;
if clk_div2=1, j:=j+1; else j:=j;
if j:=60 count_state:=01
elsif j=61, count_state:=02 else count_state:=0.
endif, end
```

Fig. 2.5: VHDL pseudocode of wind speed measurement.

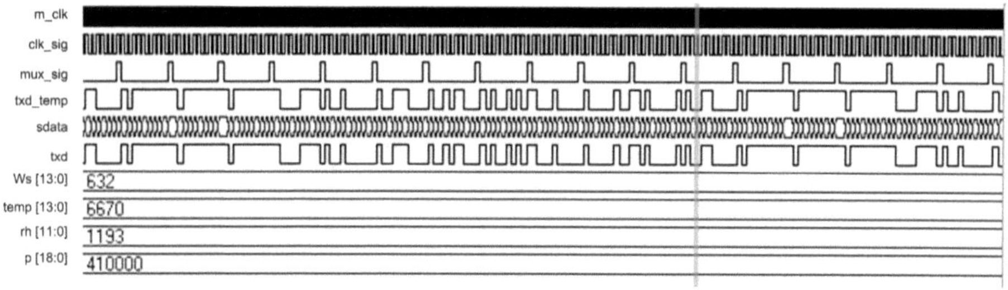

Fig. 2.6: Wind speed measurement simulated timing diagram – Modelsim results for the value $(000000010000)_2 = (16)_{10} = 2.75\ \text{m s}^{-1}$.

2.4.2 Relative humidity and temperature measurement – SHT11 Sensor

A low cost and low power surface mountable, 8 pin SHT11 sensor (Sensirion 2011) is used to measure T in 14 bit resolution and RH in 12 bit resolution. The sensor consists of a capacitive sensing element, a polymer, which absorbs and desorbs water molecules depending on the surrounding conditions and provides a fully calibrated digital output. The Two Wire Interface (TWI) and internal voltage regulation allow for fast system integration. The SHT11 is interfaced with FPGA in which a digital architecture shown in Fig. 2.7 is developed to transfer the communication/control sequence to the sensor. The digital architecture logical design encompasses two distinct parts: (i) the data path processor unit performing the data processing operations consists of shift registers, multiplexers, counters, flip-flops, demultiplexers and tristate switchs, etc. and (ii) control engine-finite state machine, that sends commands to the data path processing unit to determine the sequence in which various actions like sending the sequence of connection reset, transmission start, address and command etc. are performed.

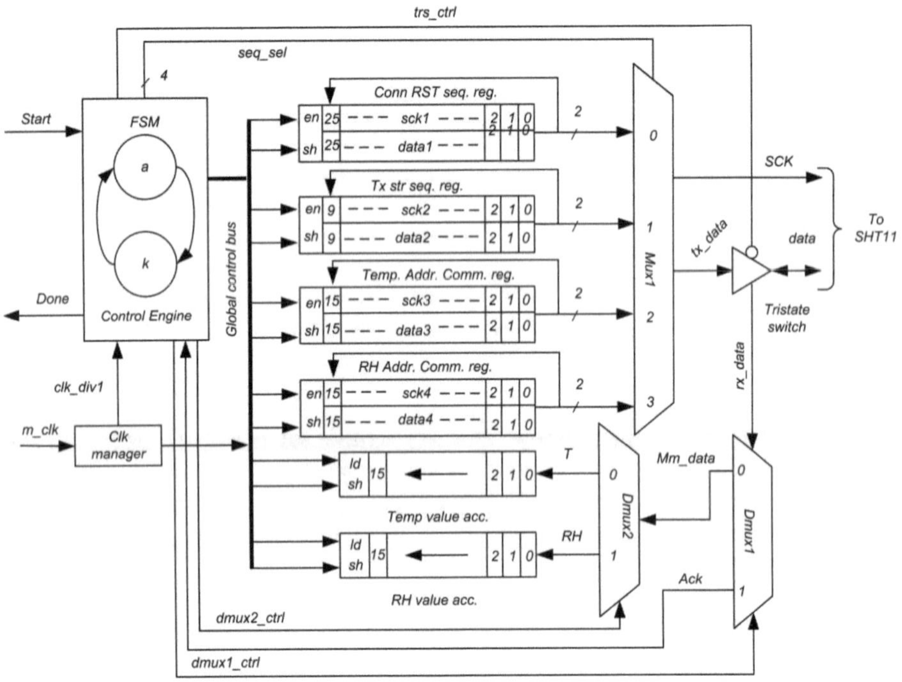

Fig. 2.7: Digital architecture for temperature and relative humidity measurement-TWI interfacing protocol.

The serial clock (SCK) and data pins are accessed by architecture and the SHT11 sensor for internal register programming, measurement data acquisition, reception of acknowledgement and mode selection etc. SCK is used to synchronize the communication between architecture and sensor. Since the interface consists of fully static logic, there is no minimum SCK frequency. The data pin is used to transfer data in and out of the sensor. The output SCK and inout data are connected to the SHT11 and are used for sending the sequence and reading the measurement values as per the execution command from the control engine. The control engine state transition flow occurs from state 'a' through state 'k' for every T and RH measurement cycle. The signal dmux1_ctrl is used to control the Dmux1 to transfer the rx_data either to Dmux2 (Mm_data) or to the control engine (ACK). This operation sequence is repeated to collect the T and RH data over a long period. The data shifting, circular rotations, enabling sequence register, and data loading at the desired accumulator are performed by the control engine, multiplexer and demultiplexer. The tristate switch output is inout, i.e. it is driven by the input when enabled otherwise driven by the SHT11. The timing diagram synchronization among all the operations of the measurement is maintained by the clock manager. The control engine state transition flow for T and RH measurement is shown in Fig. 2.8 and explained below:

State (a): The connection reset sequence is stored in a [25:0] circular shift register such as to toggle SCK nine times during data high followed by toggling SCK one time at data low and forcing SCK and data high again low. This sequence resets the status register of SHT11 with the default contents.

State (b): The transmission start sequence is stored in a [9:0] circular shift register such as to lower the data while SCK high followed by a low pulse and rising data high while SCK high.

State (c): The temperature measurement address and command (0x03H) is stored in a [15:0] circular shift register and transferred to the SHT11 for the rising edge of the SCK.

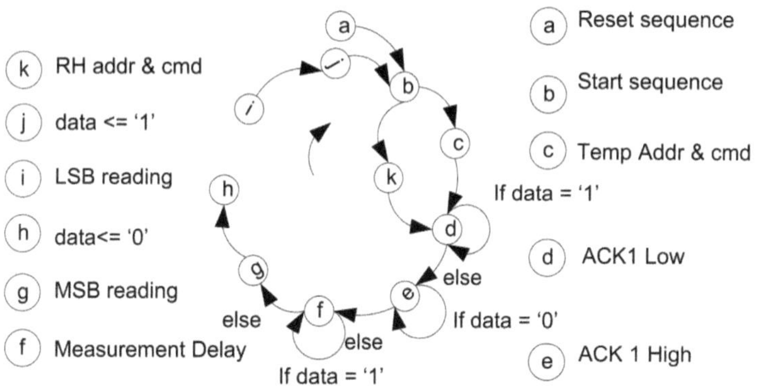

Fig. 2.8: Finite state machine control engine state transition flow for temperature and relative humidity measurement.

State (d): The SHT11 sensor indicates the proper reception of the address and command by pulling the data low after the trailing edge of the SCK (ACK1 Low).

State (e): Check for whether data is zero or not.

State (f): Delay for measurement approximately 320 ms. The completion of the measurement is signaled by pulling the data low.

State (g): Reading the MSB (D15–D8) of the measurement value bitwise and loading into the [15:8] of T value accumulator for the rising edges of SCK.

State (h): An acknowledgment is passed to the SHT11 by forcing the data low after reading MSB data followed by a pulse on SCK.

State (i): Reading the LSB (D7–D0) of the measurement value bitwise and loading into the [7:0] of T value accumulator.

State (j): Forcing the data high since the CRC is not used followed by the state (b) operation to read out the RH measurement data.

State (k): The RH measurement and command (0x05H) is stored in a [15:0] circular shift register and transferred to the SHT11 followed by state (d) to state (j) operations.

The signal dmux1_ctrl is used to control the Dmux1 to transfer the rx_data either to Dmux2 (Mm_data) or the control engine (ACK). In states (g) and (i), the RH value accumulator is enabled by Dmux2_ctrl while the relative humidity measurement is being carried out.

The digital readout T_{meas} conversion formula to calculate the equivalent temperature in °C is (Sensirion 2011)

$$\text{Temp} = T_{\text{in}\,^\circ\text{C}} = -40.1 + 0.01T_{meas}\,. \tag{2.5}$$

The digital readout RH_{meas} conversion second order formula to calculate the true RH in percentage with the temperature compensation is (Sensirion 2011)

$$\begin{aligned} RH_{\text{in}\,\%} &= (T_{\text{in}\,^\circ\text{C}} - 25)(0.01 + 0.00008RH_{meas}) \\ &\quad + 0.0367RH_{meas} - 1.5955 \cdot 10^{-6}RH_{meas}^2 - 2.0468\,. \end{aligned} \tag{2.6}$$

The due point temperature T_d is calculated from RH and T readings with the following approximation in good accuracy (Sensirion 2011)

$$T_d = 243.12\left(\frac{\ln\left(\frac{RH}{100}\right) + \left(\frac{17.72T}{243.12+T}\right)}{17.62 - \ln\left(\frac{RH}{100}\right) - \left(\frac{17.62T}{243.12+T}\right)}\right). \tag{2.7}$$

The pseudocode of measurement algorithm and simulation timing response are shown in Figs. 2.9 and 2.10 respectively.

2.4.3 Absolute pressure measurement – SCP1000-D01 sensor

The SCP1000-D01 sensor is used to measure the absolute pressure in 19 bit resolution. The sensor consists of a silicon bulk micromachined sensing element chip and a signal conditioning Application Specific Integrated Circuit (ASIC).

Overcoming the timing error, more accurate measurements can be carried out using SCP1000-D01 (Witzel 2008). The pressure sensor element and the ASIC are

Declarations: m_clk:=in:=0; sck:=out:=0;
data:=inout:=0; trs_ctrl,dmux1_ctrl, demux2_ctrl:=wire:=0;
global_ctrl_bus(en,sh,ld):=wire:=0;
variable i,j,sck5,sck6,data5,data6 :=integer:=1;
Process(1): clk_div1<=not clk_div1, i:=1 if i:=j else i:=i+1;
Process(2): data<=tx_data when trs_ctrl=0 else z; rx_data<=data;
Process(3): @clk_div1: if seq_sel_cnt is = 0, con_rest_seq_sck&dataon;
1|2,trans_st_seq_sck &data on; 2|10, (i) temp_addcom_data & sck on,
(ii) rh_addcom_data on, 3|11 trs_sck<=1:demux1_ctrl<=1, check for ack high;
4|12, check for ack low, delay (); 5|13, check for ack high, delay (), check for ack low;
6|14, demux1_ctrl <=0; (i) demux2_ctrl<=0, (ii) demux2_ctrl<=1, (i) read_temp_msbon, (ii) read_rh_msbon;
7, trs_ctrl<=0, check for ack low; 8|15, trs_ctrl<=1, (i) demux2_ctrl<=0, (ii) demux2_ctrl<=1, (i) read_temp_lsbon, (ii) read_rh_lsbon; end;
Process(4): mm_data<= rx_data if demux1_ctrl=0 else ack<=rx_data
t<=mm_data if demux2_ctrl=0 else rh<=mm_data;
Process(5): @clk_div1, case seq_sel is 0, crs_sck <= crs_sck (24:0) & crs_sck(25:0);
crs_data<=crs_data (24:0) & crs_data(25:0); 1|9, ts_sck<=ts_sck(8:0)&ts_sck(9);
ts_data<=ts_data(8:0)&ts_data(9); 2|10,
temp_add_com_sck<=temp_add_com_sck(14:0) &temp_add_com_sck(15); (i)
temp_add_com_data<=temp_add_com_data(14:0)& temp_add_com_data(15), (ii)
rh_add_com_data<= rh_add_com_data(14:0)& rh_add_com_data (15); 4,
ack_sck<=ack_sck(1:0)&ack_sck(2);
6|12, rd_sck<=rd_sck(14:0)&rd_sck(15); 7|13, rd_ack_sck<=rd_ack_sck (2:0)&
rd_ack_sck (3); rd_ack_data<=rd_ack_data (2:0) & rd_ack_sck (3);
8|14, rd_sck<=rd_sck (14:0) & rd_sck(15), end case;
Process(6): if seq_sel is 0=>sck<=sck1:tx_data<=data1; 1|9=>sck<=sck2:tx_data
<=data2; 2|10=>sck<=sck3:tx_data<=data3; 4=>sck<=sck4; 6|8|12|14=>sck<=sck5;
7|13=>sck<=sck6:tx_data<=data6; end;

Fig. 2.9: VHDL pseudocode of RH and T measurement.

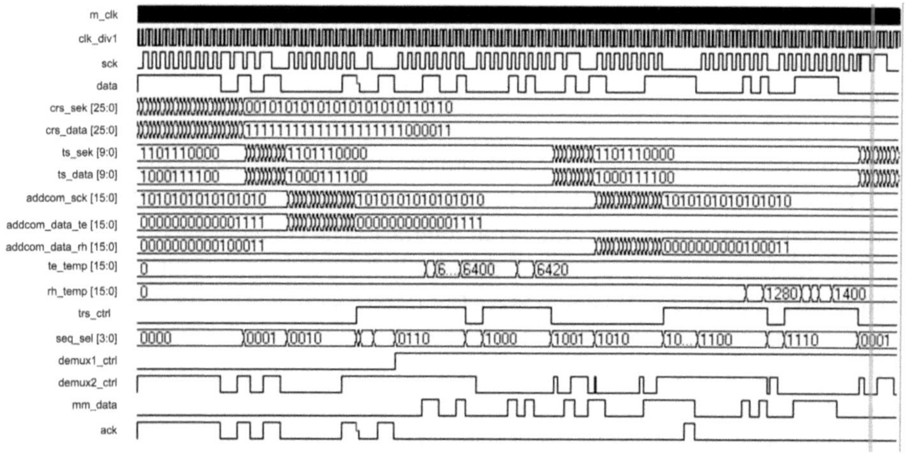

Fig. 2.10: Timing diagram of T and RH measurement simulated in Modelsim for the values 01100100010100 = $(6420)_{10}$ = 24.10 °C and 010101111000 = $(1400)_{10}$ = 46.09 %.

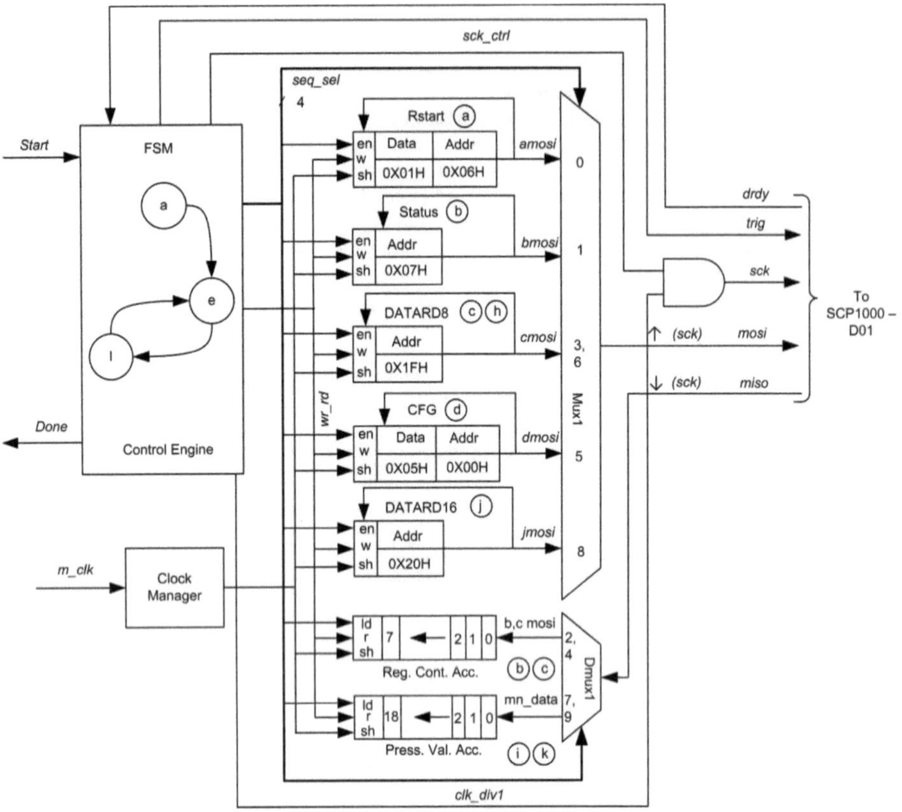

Fig. 2.11: Digital architecture of pressure measurement: SPI communication interface protocol.

mounted inside a plastic pre-mold package and wire bonded to appropriate contacts. The pressure output data are calibrated and compensated internally. The digital architecture of pressure measurement is shown in Fig. 2.11. The pressure measurement digital architecture consists of two parts: one is a data processor unit and the other is the control engine. The communication protocol between the digital architecture and SCP1000-D01 is a Serial Peripheral Interface (SPI) (VTI Technologies 2007). The SPI interface is a full duplex five wire serial interface. The communication between the architecture and sensor is done with data ready (DRDY), trigger (TRIG), SCK, Master Out Slave In (MOSI) and Master In Slave Out (MISO) signals. The SPI communication frame consists of three 8 bit words. The first word defines the register address followed by the type of access, i.e. '0' for read and '1' for write and one '0' at LSB followed by the data words being read or written. The MSB of the words are sent first. Bits from the MOSI line are sampled for the rising edges of SCK while bits to the MISO are latched out for the trailing edge of SCK. The register address and data sequence are stored in the circular shift registers in the data processor unit. The state transition cycle of finite-state-machine control engine corresponding to the pressure measurement occurs every second. The register address, content sequence shifting, enabling the register, loading the content, writing and reading are performed at the data processor unit as per the command sequence from the control engine. The SCK_ctrl controls issuing of clk_div1 to the SCK. The timing synchronization among various operations of the whole measurement is maintained by the clock manager. The register address and data sequence are stored in the circular shift registers in the data processor unit. The FSM control engine state transition flow for pressure measurement is shown in Fig. 2.12 and explained below:

State (a): The restart register (Add:0x06H) is written with restart sequence (data: 0x01H).

State (b): Reading the content of status register(Add:0x07H) and loading bitwise into the register content accumulator followed by checking the Least Significant Bit (LSB) to verify that startup procedure is finished.

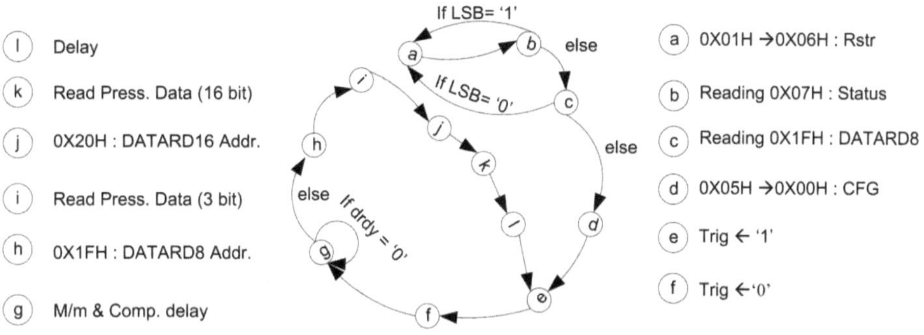

Fig. 2.12: Finite state machine control engine state transition flow for pressure measurement in triggered mode.

State (c): Reading the content of datard8 register (Add:0x1FH) and loading bitwise into the register content accumulator followed by checking the LSB to identify that the SCP100-D01 is standby mode and waiting for measurement command.

State (d): The configuration register (Add:0x00H) is written with 17 bit measurement sequence (data:0x05H).

State (e): Forcing the trigger signal high.

State (f): Forcing the trigger signal low.

State (g): Wait for measurement and computation delay till the data ready goes high.

State (h): Writing the datard8 (Add:0x1FH) address sequence.

State (i): Reading the content of datard8 register and bitwise loading into pressure value accumulator.

State (j): Writing the datard16 (0x20H) address sequence.

State (k): Reading the content of datard16 register and bitwise loading into pressure value accumulator.

State (l): Wait for delay to start next measurement cycle.

The pseudocode of measurement algorithm and simulation timing response are shown in Figs. 2.13 and 2.14 respectively.

```
Declarations: m_clk, miso:=in:=0;
drdy:=in:=1; sck, mosi, trig:=out:=0;
clk_div1, wr_rd:=wire:=0; sck_ctrl:=wire:=1;
amosi[17:0], bmosi[8:0], cmosi[8:0], dmosi[16:0],
hmosi[8:0]:=wire:=value;seq_sel[3:0]:=wire:=0;
bmiso[7:0], cmiso[7:0],Mm_data[18:0]:=wire:=0;
i,j,k:=variable:=integer:=1;
process(1): while(1)
clk_div1<=not clk_div, i:=0 if i:=k; else i:=i+1;
Process(2): if cnt:=0, wr_rd<= 0: sck_ctrl<= 1;
elsif cnt:=2, wr_rd<=1, delay (), wr_rd<=0;
elsif cnt:=3,if lsb=0, cnt:=cnt+1; else cnt:=0;
elsif cnt:=4 | 9 | 11, wr_rd<=1;
elsif cnt:=5, wr_rd<=0, sck_ctrl<=0;
elsif cnt:=6, if lsb=1, cnt:=cnt+1, else cnt:=0;
elsif cnt:=7, sck_ctrl<=1;trig<=1, delay(); trig<=0;
elsif cnt:=8, if drdy<=0; cnt:=8;else cnt:=cnt+1;
elsif cnt:=10|12, wr_rd<=0;
Process (3): case seq_sel is
if wr_rd=0, when 0:=>mosi<=amosi(n-1:0)&amosi(n);
when 1:=>mosi<=bmosi(n-1:0)&bmosi(n);
when3|8:=>mosi <=cmosi(n-1:0)&cmosi(n);
when6: =>mosi<=dmosi(n-1:0)&dmosi(n);
when10: =>mosi<=hmosi(n-1:0)&hmosi(n);
else when2: =>bmosi<=bmosi(n-1:0)&miso;
when4: =>cmosi<=cmosi(n-1:0)&miso;
when5|6: =>Mm_data<=Mm_data(n-1:0)&miso; end;
```

Fig. 2.13: VHDL pseudocode of P measurement.

The transmission goes to state (e) after finishing the first measurement cycle and continues collecting the pressure data over a long period. The register address, content sequence shifting, enabling the register, loading the content, writing and reading are performed at the data processor unit as per the command sequence from the control engine.

The SCK_ctrl controls issuing of clk_div1 to the SCK. The true pressure value Pr is calculated in kPa by (VTI Technologies 2007)

$$\mathrm{Pr} = 0.25(\mathrm{P_{meas}})_{10} \,. \tag{2.8}$$

The measurement starts when the ‘Start’ pin is high and ‘Done’ goes high then low to signal the completion of every measurement cycle.

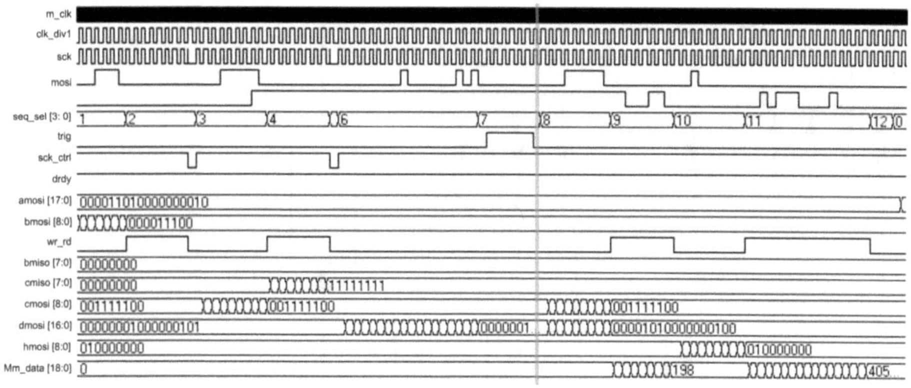

Fig. 2.14: Timing diagram of P measurement simulated in Modelsim for the value $(1100010111000010000)_2 = (405008)_{10} = 101.252$ kPa.

2.5 Communication protocol and frame format

The meteorological data acquisition measurement is done every second and the values are stored in radix-2 format in the specified registers. These data are continuously shifted to the M/m data buffer during subsequent measurements without data conflict/loss. The measured data are plunged into the communication architecture and propagated through the modules corresponding to Universal Asynchronous Receiver Transmitter (UART)–RS232 communication protocol to the data logging computer (Nhivekar & Mudholker 2011). The UART–RS232 standard serial communication protocol (Cheever 2010; Esposito et al. 2011) is implemented as a separate digital architecture inside the FPGA as shown in Fig. 2.15 (a). The contents of the buffer are shifted into the communication frame planer for every rising edge of data_rd signal. The communication frame planer splits the measurement data intact length into bytes, introduces start bit ('0') before every byte and concatenates sufficient zeros with the bits, as shown in Fig. 2.15 (b), to form the communication frame and sends to mux1.

In order to avoid the occurrence of communication conflict among the frames, two header (H1 and H2) frames are transferred followed by the measurements data frames and one footer (F1) frame as an appropriate coordination among the data, so that totally 12 frames are transferred for every measurement cycle. The count value of cnt2 informs the byte shift controller about the completion of one frame transmission (i.e. 11 bit) and then the byte shift controller increments cnt3 to select the next input (frame) of mux1 to load it into the Parallel In Serial Out (PISO) shift register. The baud rate counter generates the bits transmission clock i.e. baud rate clock clk_sig at the rate of 9600 from the master clock m_clk. The shift register performs a cyclic shift on its content sdata (sdata = frame format of H1, H2, Ws, T, RH, Pr, F1) for the rising edges

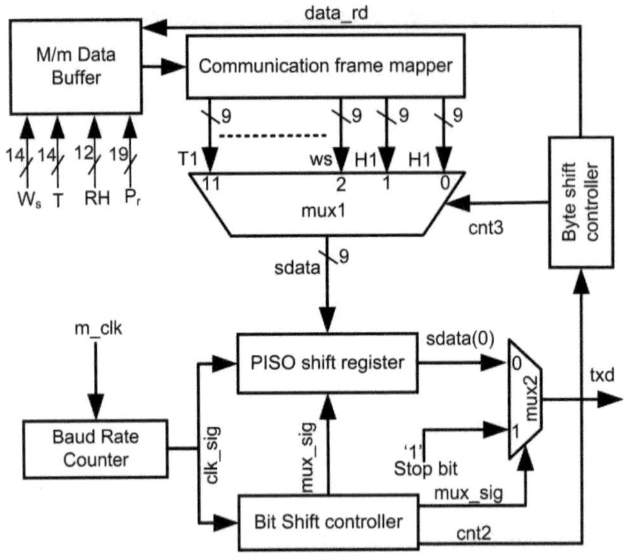

(a)

cnt3:=0,1 H1,H2
Cnt3:=2,3 Ws[7:0],"00"&Ws[13:8]
Cnt3:=4,5 Temp[7:0],"00"&Temp[13:8]
Cnt3:=6,7 Rh[7:0],"0000"&Rh[11:8]
Cnt3:=8,9,10 Pr[7:0],Pr[15:8],"00000"&Pr[18:16]
Cnt3:=11 T1

(b)

Fig. 2.15: (a) UART–RS232 communication protocol architecture and (b) communication frame format.

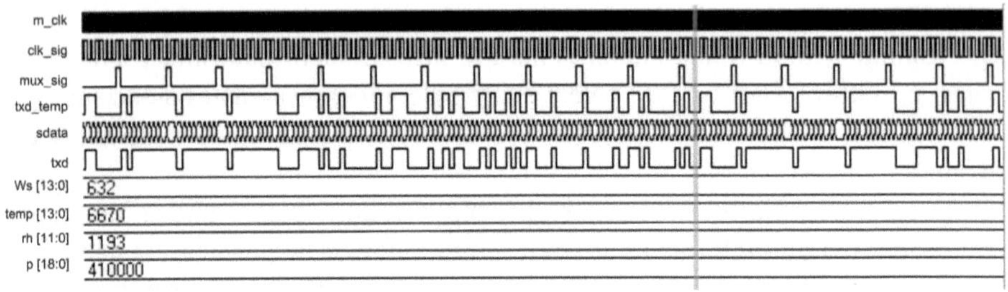

Fig. 2.16: Timing diagram of UART communication protocol simulated in Modelsim for Ws = 00001001111000 = $(632)_{10}$ = 98.2458 m s^{-1}, T = 1101000001110 = $(6670)_{10}$ = 26.6 °C, RH = 010010101001 = $(1193)_{10}$ = 39.63 %, and P = 1100100000110010000 = $(410000)_{10}$ = 102.5 kPa.

of the transmission clock and the sdata (0) is forwarded to the first input of the mux2 when the mux_sig is high, otherwise the shift register is disabled. The timing diagram and simulation result of the above operation are shown in Fig. 2.16.

The bit shift controller maintains mux_sig at high for the entire duration of the frame transmission and pulls back to low until loading the next frame into the PISO shift register and continues the similar transitions. The mux2 sends sdata(0) to the data logging computer when mux_sig is high, otherwise it sends stop bit '1'.

2.6 Performance calibration of the proposed measurement system

The accuracy, repeatability, reproducibility and uncertainty (Kušnerovál et al. 2013; Paulsen et al. 2007) of the proposed measurement results were tested through extensive experiments. The proposed system can measure the wind speed from 0.27 m s^{-1} to 60 m s^{-1} with 0.25 m s^{-1} resolution; temperature from −40 °C to 100 °C with 0.01 °C resolution; relative humidity from 0 % to 100 % with 0.05 % resolution and pressure from 30 kPa to 120 kPa with 1.5 Pa resolution (Pena & Duro 2003; Sensirion 2011; VTI Technologies 2007). However, the final accuracy of the measurement depends on sensor precision and linearity over the periods of the experiment. The measured data are compared against the standard (Lutron) measurement instruments and uncertainty in measurements is estimated using the Type A and Type B guidelines of the National Accreditation Board for Testing and Calibration Laboratories (NABL) (NABL 2000) and the National Physical Laboratory (NPL) (Cox & Harris 2006) as given by

$$\text{Type A:} \quad U(\overline{x}) = \sqrt{\frac{1}{n}\frac{1}{n-1}\sum_{i=1}^{n}(x_i - \overline{x})^2} \tag{2.9}$$

where $U(\overline{x})$ is standard uncertainty (error mean), $\overline{x}$ is the arithmetic mean value (best measurement), n is number of samples, x_i is measurement elements

$$\text{Type B:} \quad Uc = \sqrt{U(\overline{x})^2 + U_{b1}^2 + U_{b2}^2 + U_{b3}^2 + U_{b4}^2} \tag{2.10}$$

where U_{b1}^2 is master instrument uncertainty, U_{b2}^2 is accuracy of the master instrument, U_{b3}^2 is resolution of the test instrument and U_{b4}^2 is instrument error.

The subsets of proposed and standard measurement data samples with 2000 elements obtained from the experiments are shown in Fig. 2.17 and a good level of stability as well as accuracy between the proposed and standard measurements are observed. The minimum values of average errors are observed due to the tolerance of the sensor and errors introduced by the internal Analog to Digital converter (A/D) (Wekesa et al. 2013). The wind speed and pressure measurement graphs in Fig. 2.17 (a) and (d) show very good accuracy for the standard measuring instrument except for a few data. From the graph of temperature and relative humidity in Fig. 2.17 (b) and (c) it

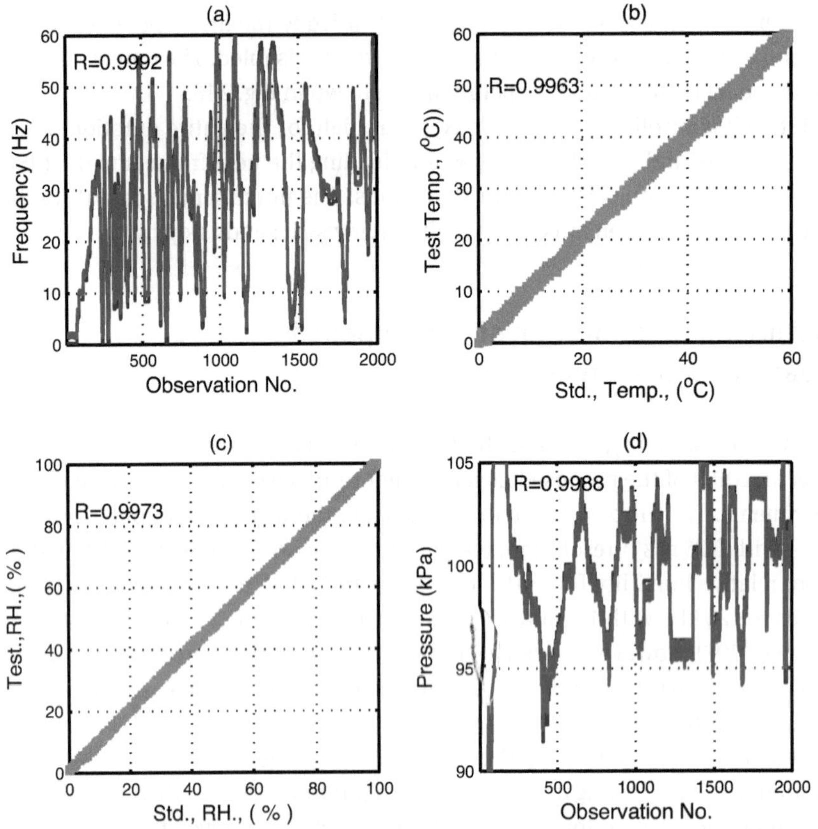

Fig. 2.17: Comparison between the standard and proposed measurement data for ascending and descending input variations: standard (blue), test (red), linear regression (black) and residual (green).

is clearly seen that there is very close agreement between the proposed and standard measurement system around 40 °C for T and 50 % for RH and the slight deviations arise with increasing/decreasing inputs with maximum average errors of ±1 °C and ±2 % respectively due to the nonlinearity of the sensors. The highest correlation coefficients of $R = 0.9992$ and $R = 0.9988$ are achieved for wind speed and pressure measurements and the correlation coefficients of $R = 0.9963$ and $R = 0.9973$ are obtained for temperature and relative humidity measurements respectively.

The gains, constants and conversion coefficients used in Eqs. (2.4)–(2.8) are obtained from the calibration experiments and the sensor manufacturer data sheets to the maximum measurement accuracy (Andria et al. 2005). Regression analyses for repeatability and reproducibility are carried out for different reference setpoints. During these tests a set of ten measured samples are collected from the proposed and standard measurement system for every setpoint variation and hence each test point yields

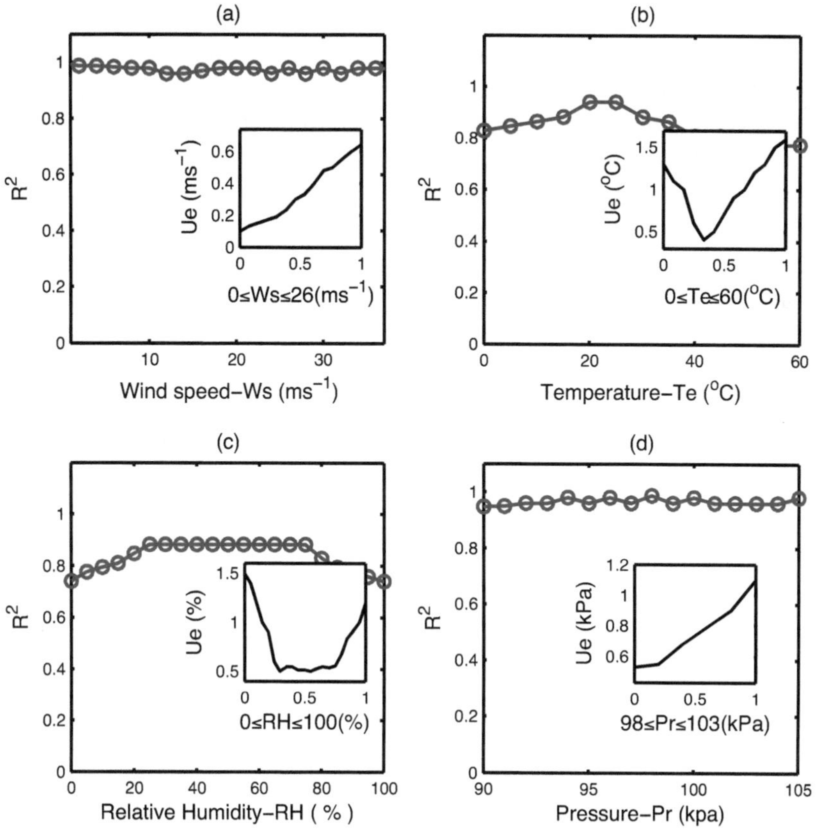

Fig. 2.18: Performance plots of coefficient of determination (red): (a) wind velocity, (b) relative humidity, (c) temperature, and (d) pressure. The inset figures (black) show the achieved accuracy of measurement uncertainty (U_e) against the respective normalized environmental parameter.

10×2 matrices. The statistical computations of the coefficient of determination R^2 are carried out for the elements of the matrices and the results are used to represent the degree of linearity between the proposed and standard measurements. Figure 2.18 shows the performance plots of R^2 for wind speed, temperature, relative humidity and pressure. From Fig. 2.18 (a) and (d) it is observed that the variations of the coefficients of determination for the wind speed and pressure are approximately close to unity which shows that the proposed and standard measurement data fit exactly each other with very good linear relations. Further, the least values of R^2 are 0.982 and 0.973 respectively throughout the variations of the reference setpoints. From Fig. 2.18 (b) and (c) it is observed that (i) R^2 appears below unity in some regimes, (ii) slight nonlinearity exists between the proposed and standard measurement, (iii) increasing and decreasing trends are seen in R^2 with increasing setpoint inputs, (iv) R^2 reaches a maximum at 25 °C and goes down with increasing/decreasing the reference setpoints, and (v) in contradiction to temperature, the R^2 values for relative humidity increase until the reference setpoint reaches 30 %, is constant from 31 % to 80 % and decreases for the remaining values of setpoints. The minimum and maximum R^2 values of temperature and relative humidity are (0.8812, 0.9723) and (0.8661, 0.9418) respectively. However, this resolution and accuracy are more sufficient for various applications including the estimation of low altitude atmospheric turbulence strength (C_n^2). The expanded uncertainty (U_e) is calculated using Eqs. (2.9) and (2.10) which increases the probability dispersion to 95 % and thus the reliability of declared values as well (Kušnerovál et al. 2013). The effective degrees of freedom are infinite and hence the expanded uncertainty (U_e) is calculated with the combined uncertainty (Pena & Duro 2003; Coquilla 2010; NABL 2000) and the coverage factor 'k' is equal to 2 at the confidence level of 95.45 %. The figures in the box in Fig. 2.18 show the expanded uncertainty of the proposed measurement system against the normalized inputs. The input ranges are normalized to the typical local atmospheric parameter fluctuations in outer scale and given in the xlabel of the figures in the box. The inset figures clearly exhibit the exactness and capability of the proposed measurement system.

2.7 Atmospheric turbulence strength (C_n^2) estimation

Although several dozen turbulence profile models have been developed from the experimental measurements made at a variety of locations, no model provides generalization. Most of the models (SLC-day, Hufnagel–Vally night and Greenwood etc.) are a function of altitude (h), i.e. vertical path, which is unfit for C_n^2 estimation for terrestrial FSOC (horizontal or slant path) and they yield reasonably good estimations only for the particular location/time, i.e. daytime, nighttime, mountaintop location, China lack field – over land and over water etc. (Majumdar & Ricklin 2008; Barrios & Dios 2012). Further, the turbulence fluctuation in the surface boundary layer not only varies as a function of altitude, but also according to local conditions such as terrain type, geographical location, cloud cover, meteorological values and local time of day. The PAMELA model provides C_n^2 estimations within the surface boundary layer and it accepts all the parameters of test field geographical location, meteorological values and optical path as the inputs. Therefore the PAMELA model is preferred and modified according to local test field parameters. A separate MATLAB code is developed to estimate the turbulence strength once in every sixty seconds.

The required geographical inputs are latitude (10°38′46.7334″, 10°38′52.8468″), longitude (79°3′12.0774″, 79°2′56.6268″), time of day (diurnal period; GMT +5.30), terrain type (0.03 m open flat terrain, grass, few isolated obstacles), number of days (as applicable), height above the ground (15.25 m), and meteorological parameters at the desired altitude (15.25 m) of the experimentation for estimating the strength of C_n^2 (Tunick et al. 2005; Oh et al. 2004a; Oh et al. 2004b). The measured meteorological parameters are subsequently entered into the software (MATLAB code) for estimating the turbulence strength and subsequently updating the real-time plot and data logging table. The PAMELA model Mathcad version and its simplified form can be found in (Majumdar & Ricklin, 2008; Doss-Hammel, 2004) and the background for estimating C_n^2 are discussed in this section. The estimated solar irradiance R is used to determine the radiation class $c_r = R/300$. For a wind speed Ws define the wind speed class $c_w = \{0.27 \text{ if } Ws \leq 0.27 \text{ else } Ws\}$ and then the Pasquill stability category P can be determined by

$$P = \frac{-(4 - c_w + c_r)}{2}\,. \tag{2.11}$$

The surface roughness length for the open flat terrain, grass and few isolated obstacles is estimated from tables, $z_r = 0.03$ m, and from this it is possible to calculate the Obukhov buoyancy length scale bl

$$bl = \left[(a_1 P + a_2 P^3) z_r^{-(a_3 - a_4|P| + a_5 P^2)}\right]^{-1}, \tag{2.12}$$

where $a_1 = 0.004349$, $a_2 = 0.003724$, $a_3 = 0.5034$, $a_4 = 0.231$, $a_5 = 0.0325$. The mean vertical velocity W and fluctuating part w, horizontal velocity U and fluctuating part u define vertical momentum flux in terms of the eddy viscosity K_m and mean potential temperature Θ and fluctuating part θ define vertical heat flux in terms of the eddy diffusivity of heat K_h by

$$\overline{uw} = -K_m\left(\frac{\partial U}{\partial Z}\right) \quad \text{and} \quad \overline{\theta w} = -K_h\left(\frac{\partial \Theta}{\partial z}\right). \tag{2.13}$$

The mean specific humidity Q and fluctuating part q define the vertical water vapor flux using the eddy diffusivity of water vapor K_w by

$$\overline{qw} = -K_w\left(\frac{\partial Q}{\partial Z}\right). \tag{2.14}$$

The dimensionless wind shear $\phi_m(\zeta)$ and the dimensionless potential temperature gradient $\phi_h(\zeta)$ are expressed as functions of the scaled buoyancy parameter $\zeta = z/L$. The turbulent exchange coefficients for heat K_h and momentum K_m are given by

$$K_h = \frac{ku_* z}{\phi_h(\zeta)} \quad \text{and} \quad K_m = \frac{ku_* z}{\phi_m(\zeta)}, \tag{2.15}$$

where $k \cong 0.4$ is von Karman's constant. $K_h = K_m$ as per the optical turbulence model for laser propagation and imaging applications (Doss-Hammel et al. 2004). The friction velocity u_* and characteristic temperature T_* from the wind speed Ws and the roughness length z_r, heat flux H, specific heat c_p, and mass density ρ are given by

$$u_* = \frac{kWs}{\ln(z/Z_r)} \quad \text{and} \quad T_* = \frac{kWs}{c_p \rho u_*}. \tag{2.16}$$

The atmospheric refractive index (n) in terms of pressure (Pr) and temperature (T) is

$$n - 1 = \frac{77.6 \cdot 10^{-6} Pr}{T}\left(1 + \frac{7.52 \cdot 10^{-3}}{\lambda^2}\right) \quad \text{and} \quad \frac{dn}{dz} = -\frac{77.6 \cdot 10^{-6} Pr T_* \phi_h(\zeta)}{0.4 z T^2}. \tag{2.17}$$

The eddy dissipation rate ε and C_n^2 are estimated with the constant b ≈ 2.8 as

$$\varepsilon = \frac{u_*^3(\phi_m - \zeta)}{0.4z} \quad \text{and} \quad C_n^2 = \frac{2.8K_h}{\varepsilon^{\frac{1}{3}}}\left(\frac{dn}{dz}\right)^2 . \tag{2.18}$$

The direct relationship can be seen by expanding Eqs. (2.10) to (2.12) as

$$C_n^2 = 5.152\phi_h\left(\frac{1}{\phi_m - \zeta}\right)^{0.33}\left(\frac{77.6\cdot 10^{-6}Pr}{T^2}\right)^2 h^{-0.667}\left(\frac{-H}{C_p\rho u_*}\right)^2 . \tag{2.19}$$

Variations in the signals C_n^2 and T_* are produced by the fluctuations in the values of heat flux H and u_* as in Eq. (2.11). As can be seen from Eqs. (2.11) and (2.13), $C_n^2 \to \infty$ as wind-speed Ws → 0, so the minimum wind speed must be bounded away from zero. Further, when there is little or no wind, there is little or no turbulence, i.e. wind is required to mix the temperature gradient and create turbulence (Jurado et al. 2006a). Therefore, setting turbulence at a very low number or the threshold wind speed is the possible solution in this situation which does not introduce much inaccuracy. As per the wind sensor's manufacturer data sheet, the threshold wind speed is taken as 0.27 m s^{-1} and all computations are carried out based on this specification as suggested in Oh et al. (2004a) and Doss-Hammel et al. (2004). Since the testing of developed low cost measurement system accuracy and the correlation of PAMELA model estimation with the measured data are intensions of this work, the PAMELA model is preferred and used.

2.8 Experimental results and discussions

The meteorological profile data are acquired using the proposed measurement system. The outputs of the atmospheric sensors are of radix-2 format of different length. The sensor interfacing architectures transfer the measured atmospheric data framewise (11 bits) to the serial port of the computer where the MATLAB program reads the frames of atmospheric data every second. The frame data are manipulated to obtain the corresponding real-time changes of true values of atmospheric parameters. The corresponding plots and data recording work sheet file (.xls/.doc) in real time are updated every second. The low-altitude atmospheric turbulent strength C_n^2 is estimated using the PAMELA model from the extracted true values for statistical analysis on the weather and turbulence strength profiles for different environmental conditions. The accuracy of the estimation is validated by comparing the estimated C_n^2 with the measured C_n^2, i.e. using scintillometer data and the Rytov method as discussed in Section 2.3. A sample data logging .xls file worksheet generated on 17/04/2013 (Wednesday) is given in Tab. 2.2.

A record consisting of minimum and maximum values of the meteorological parameters and atmospheric turbulence strength is prepared at the end of the every measurement day and is shown in Tab. 2.3.

Tab. 2.2: A portion of measured meteorological data recorded in an Excel work sheet.

Observation ID	Sample Data and Time	Wind Speed ($m\,s^{-1}$)	Temperature (°C)	Relative Humidity (%)	Pressure (kPa)
41406	4/17/2013 11:30:03	2.2	37	38	100.6
41407	4/17/2013 11:30:04	2.4	36	37	100.6
41408	4/17/2013 11:30:05	2.3	37.4	38	100.6
41409	4/17/2013 11:30:06	2.5	37	38	100.6
41410	4/17/2013 11:30:07	2.3	36	37.6	100.6

Tab. 2.3: A portion of a daywise local meteorological data and estimated optical turbulence strength record.

Date (2013)	Ws ($m\,s^{-1}$)		T (°C)		RH (%)		Pr (kPa)		C_n^2 ($m^{-2/3}$)		Comment
	Min	Max	Min	Max	Min	Max	Min	Max	Min	Max	
12 Feb	1	1.5	29	32	77	84	100.5	100.6	5.63E−17	2.26E−16	Clear
22 Feb	13	15	42	45	30	60	100.4	100.8	9.09E−16	1.44E−15	Moderate
07 Mar	2.10	3.11	29	30	70	84	100.5	100.6	1.65E−14	1.16E−13	Weak
27 Mar	2	3	16	19	45	48	100.0	101.0	1.51E−13	5.87E−13	Strong

The local environmental data acquired for a diurnal period is recorded and the environmental change patterns are studied. Investigation of the validation of estimated C_n^2 with the scintillometer readings is carried out and the accuracy of the estimation in terms of correlation is analyzed. Recordings of real-time meteorological profile data, variation levels and estimation and measurement analysis of optical turbulence C_n^2 for five days in different local seasons, i.e. winter, presummer, summer, monsoon and rainy (since C_n^2 is mainly seasonally dependent) are described in this section.

2.8.1 Data for 28th December 2012, winter

The diurnal behavior of the atmosphere in real-time updated plots corresponding to the weather history profile for the whole day from MN to Nn to MN is shown in Fig. 2.19 (a–d). The variations from 0.27 $m\,s^{-1}$ to 3.528 $m\,s^{-1}$ with standard deviation (SD) of 0.959 $m\,s^{-1}$ for wind speed, 22 °C to 26 °C with SD of 1 °C for temperature, 61 % to 94 % with SD of 7 % for relative humidity and 100.7 kPa to 101.1 kPa with SD of 0.112 kPa for barometric pressure are observed from Fig. 2.19 (a–d). The different atmospheric conditions observed on 28/12/2012 (Friday) are mostly cloudy, hazy, light drizzle, overcast and misty. The SD of Ws, T, and RH are significantly low while Pr is very low.

The experimental local time series plot of C_n^2 data corresponding to the local meteorological data collected on 28th December 2012 proved that the estimated value

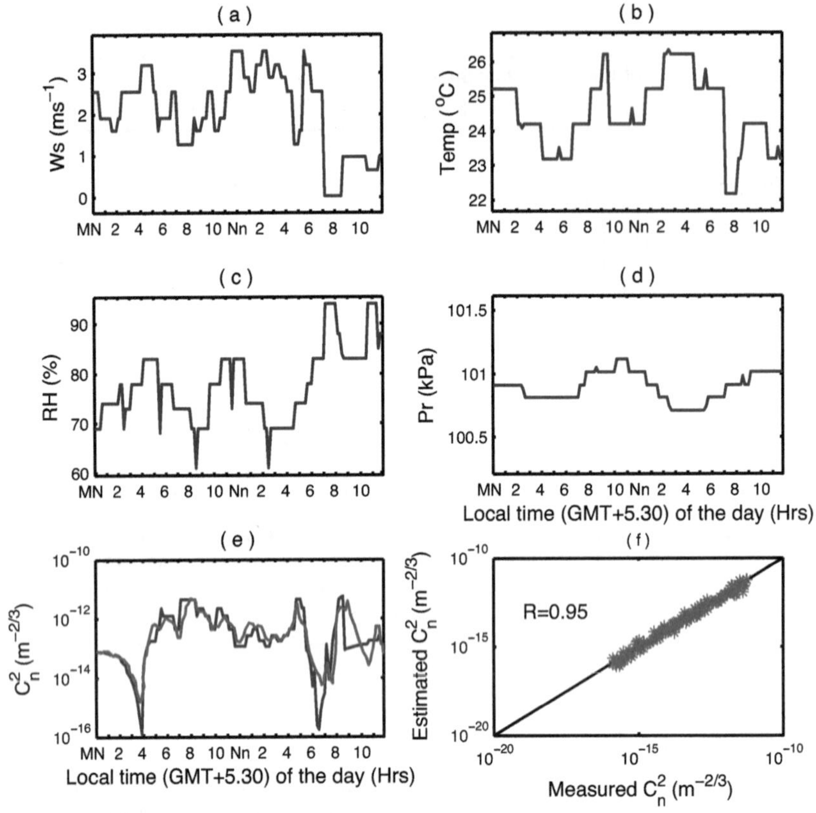

Fig. 2.19: Diurnal time series profile of meteorological parameters over a one day period. (a–d) A comparison of time series plots of atmospheric turbulence strength (C_n^2) estimated using PAMELA model (blue) and direct transmission measurement using scintillometer setup, i.e. based on signal intensity ensemble average data (red). (e) Correlation statistic plot for measured and estimated C_n^2 values. (f) MN: midnight and Nn: noon.

could yield an approximate close correlation to the measured values at almost all times of day as shown in the Fig. 2.19 (e). The measurement analysis reveals that the correlation coefficient (R) flicks in the inner scale between 95 % and 97 % and the coefficient of determination (R^2) between 90 % and 94 %. Figure 2.19 (e) shows a low turbulence intensity $\approx 1.246 \cdot 10^{-16}\,m^{-2/3}$ and $1.726 \cdot 10^{-16}\,m^{-2/3}$ about 3.55 am and 6.50 pm respectively. A high turbulence intensity $\approx 4.455 \cdot 10^{-12}\,m^{-2/3}$ is seen in the daytime since a great fluctuation is observed in the wind speed, relative humidity, pressure, and temperature values as shown in Fig. 2.19 (a–d). The average daytime turbulence intensity is about $4.354 \cdot 10^{-13}\,m^{-2/3}$. Further, minute fluctuations are obtained in the friction velocity, characteristic temperature and C_n^2 when the wind speed oscillates around a smaller value. Typically, the stronger minima are observed before early morning and after late evening. The larger fluctuations in the meteorological values from LT 5.30 am to 8.30 pm generate unrealistically large sporadic C_n^2 values. Sustained and smooth steady variations are seen in the nighttime.

2.8.2 Data for 5th March 2013, presummer

The weather parameters vary from 0.63 m s^{-1} to 3.83 m s^{-1} with SD of 0.906 m s^{-1} for wind speed, 28 °C to 43 °C with SD of 4 °C for temperature, 22 % to 74 % with SD of 14 % for relative humidity and 100.3 kPa to 100.8 kPa with SD of 0.173 kPa for barometric pressure in the observations shown in Fig. 2.20 (a–d). The different atmospheric conditions observed on 05/03/2013 (Tuesday) are mostly cloudy, haze, light drizzle, partly cloudy and scattered clouds. The SD of Ws, T, RH, and Pr are significantly low. The experimental local time series plot of C_n^2 data corresponding to the local meteorological data collected on 5th March 2013 proved that the estimated values could yield a close correlation to the measured values at almost all times of day as shown in Fig. 2.20 (e).

The measurement analysis reveals that the correlation coefficient (R) flicks in the inner scale between 97 % and 98 % and the coefficient of determination (R^2) between 95 % and 97 %. Figure 2.20 (e) shows a low turbulence intensity $\approx 3.557 \cdot 10^{-16}$ m$^{-2/3}$ about 4.15 am and $3.047 \cdot 10^{-16}$ m$^{-2/3}$ at about 5.50 pm. A brief period of very high turbulence intensity $\approx 8.716 \cdot 10^{-12}$ m$^{-2/3}$ is seen at about 5.00 pm (late evening) since the relative humidity and pressure values are very low and the wind speed and temperature values are high as shown in Fig. 2.20 (a–d). The average daytime turbulence intensity is about $7.632 \cdot 10^{-14}$ m$^{-2/3}$. Typically, the stronger minima are observed at both ends of the day. The smooth changes in the meteorological values from LT 5.30 am to 5.45 pm generate flat variations in C_n^2 values. Even variations are seen at nighttime.

2.8.3 Data for 17th May 2013, summer

The weather parameters varied from 0.194 m s^{-1} to 3.8 m s^{-1} with SD of 0.940 m s^{-1} for wind speed, 28 °C to 43 °C with SD of 4 °C for temperature, 22 % to 74 % with SD of 14 % for relative humidity and 100.3 kPa to 100.8 kPa with SD of 0.1746 kPa for barometric pressure as shown in Fig. 2.21 (a–d). The different atmospheric conditions atmosphere observed on 17th May 2013 (Friday) are mostly hazy, partly cloudy and scattered clouds. The SD of the T and RH are significantly large while the Ws and Pr are very low. Greatly uneven wind speed variation is observed as in Fig. 2.21 (a).

The experimental local time series plot of C_n^2 data corresponding to this local meteorological data collected on 17th May 2013 proved that the estimated values could yield a very low correlation (since the PAMELA model has a high sensitivity to the Ws) to the measured values at almost all times of day as shown in Fig. 2.21 (e). The measurement analysis reveals that the correlation coefficient (R) flicks in the outer scale between 79 % and 82 % and the coefficient of determination (R^2) between 71 % and 73 %. Fig-

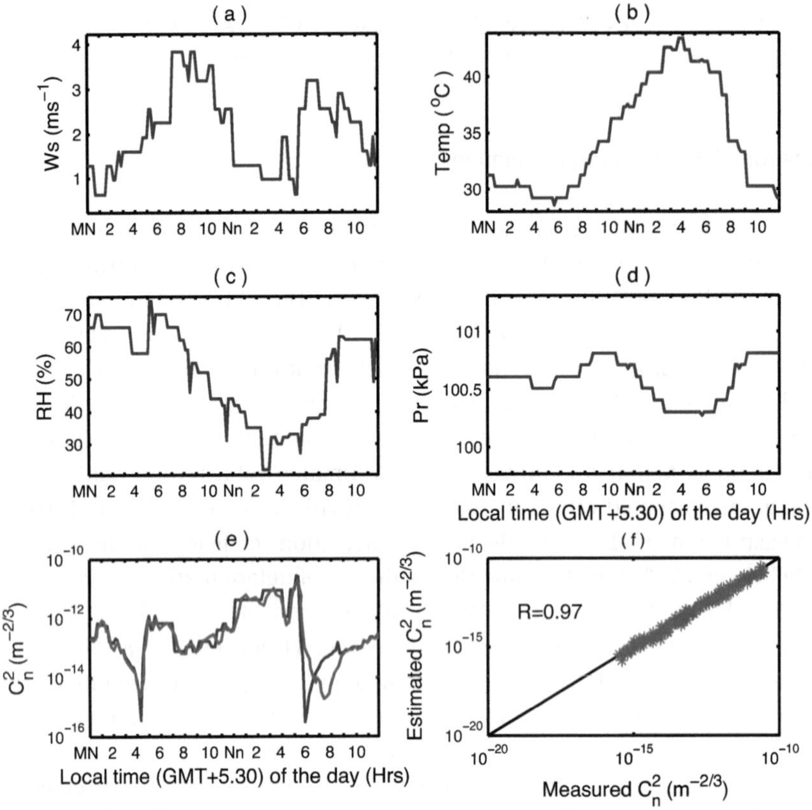

Fig. 2.20: Diurnal time series profile of meteorological parameters over a one day period. (a)–(d) A comparison of time series plots of atmospheric turbulence strength (C_n^2) estimated using PAMELA model (blue) and direct transmission measurement using scintillometer setup, i.e. based on signal intensity ensemble average data (red). (e) Correlation statistic plot for measured and estimated C_n^2 values. (f) MN: midnight and Nn: noon.

ure 2.21 (e) shows a low turbulence intensity $\approx 2.05 \cdot 10^{-16}\,\mathrm{m}^{-2/3}$ at about 5.20 pm. A greatly fluctuating turbulence intensity is seen from 4.30 am to 5.30 pm due to random fluctuations in wind speed, gradual decrease in relative humidity, low pressure and gradual increase in temperature as shown in Fig. 2.21 (a–d). The average daytime turbulence intensity is about $1.661 \cdot 10^{-13}\,\mathrm{m}^{-2/3}$. Further, minute fluctuations are obtained in the friction velocity, characteristic temperature and C_n^2 when the wind speed oscillates around a small value. Typically, the stronger minima are observed at the late evening. The larger fluctuations in the meteorological values from LT 5.30 am to 4.45 pm generate unrealistically large sporadic C_n^2 values. Sustained and smooth steady variations are seen in the nighttime.

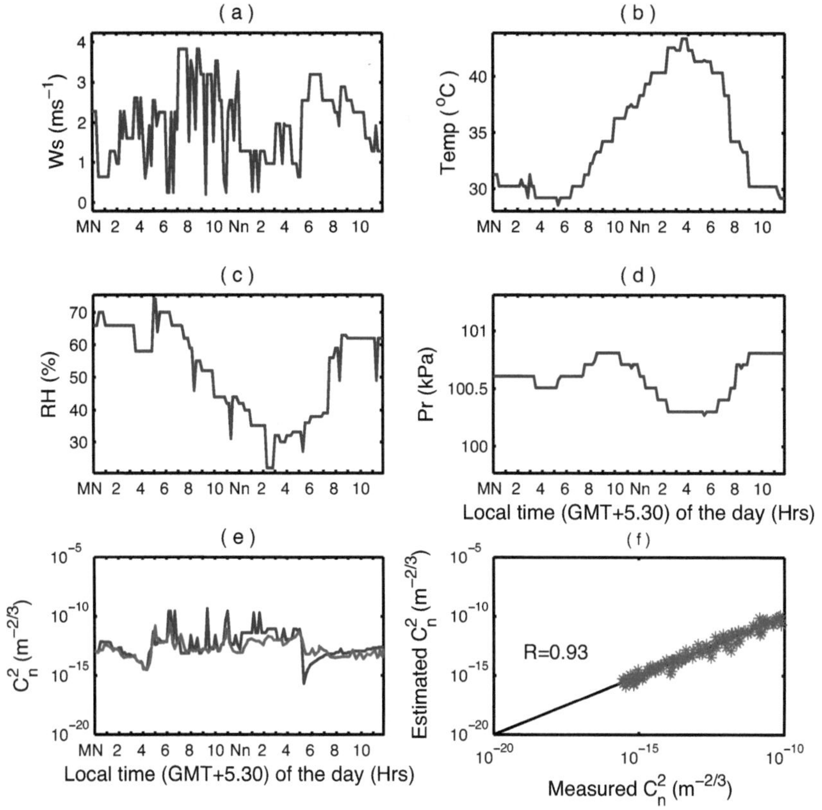

Fig. 2.21: Diurnal time series profile of meteorological parameters over a one day period. (a–d) A comparison of time series plots of atmospheric turbulence strength (C_n^2) estimated using PAMELA model (blue) and direct transmission measurement using scintillometer setup, i.e. based on signal intensity ensemble average data (red). (e) Correlation statistic plot for measured and estimated C_n^2 values. (f) MN: midnight and Nn: noon.

2.8.4 Data for 13th June 2013, monsoon

The weather parameters varied from $1\,\mathrm{m\,s^{-1}}$ to $6\,\mathrm{m\,s^{-1}}$ with SD of $1\,\mathrm{m\,s^{-1}}$ for wind speed, 28 °C to 34 °C with SD of 2 °C for temperature, 29 % to 62 % with SD of 7 % for relative humidity and 100 kPa to 100.4 kPa with SD of 0.113 kPa for barometric pressure as shown in Fig. 2.22 (a–d). The different atmospheric conditions observed on 13/06/2013 (Thursday) are overcast, cloudy and mostly hazy. The SD of Ws is high while T, RH, and Pr are low.

The experimental local time series plot of C_n^2 data corresponding to the local meteorological data collected on 13th June 2013 proved that the estimated values could yield a low correlation to the measured values at almost all times of day as shown in Fig. 2.22 (e). The measurement analysis reveals that the correlation coefficient (R)

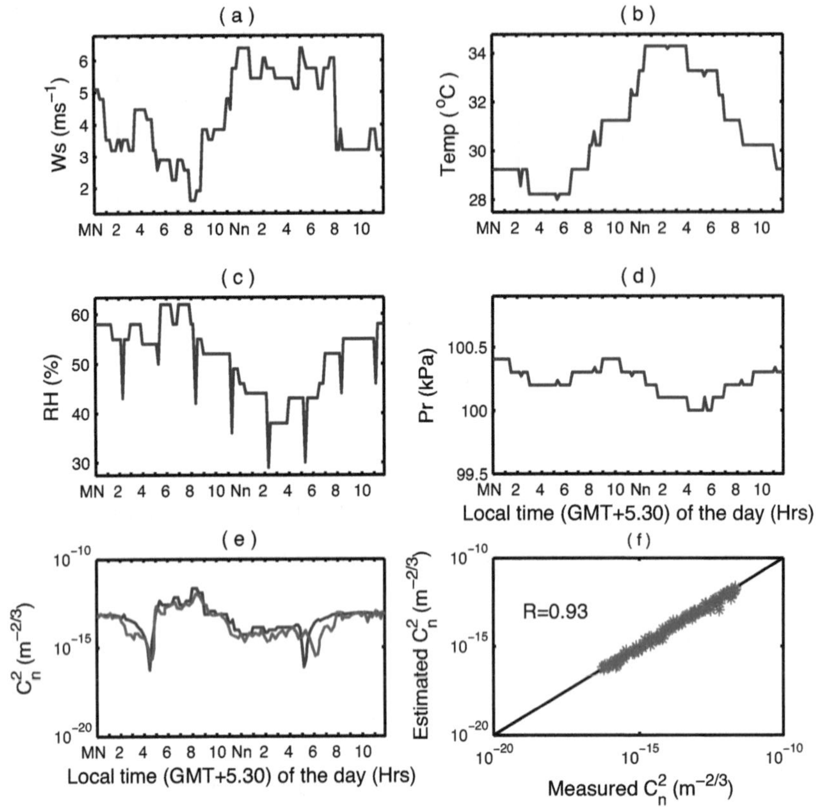

Fig. 2.22: Diurnal time series profile of meteorological parameters over a one day period. (a–d) A comparison of time series plots of atmospheric turbulence strength (C_n^2) estimated using PAMELA model (blue) and direct transmission measurement using scintillometer setup, i.e. based on signal intensity ensemble average data (red). (e) Correlation statistic plot for measured and estimated C_n^2 values. (f) MN: midnight and Nn: noon.

flicks in the inner scale between 89 % and 91 % and the coefficient of determination (R^2) between 86 % and 88 %. Figure 2.22 (e) shows a very low turbulence intensity $\approx 5.538 \cdot 10^{-17}$ m$^{-2/3}$ and $7.837 \cdot 10^{-17}$ m$^{-2/3}$ at about 4.50 am and 5.30 pm respectively. A few short intervals of low turbulence intensity $\approx 9.654 \cdot 10^{-15}$ m$^{-2/3}$ are seen about noon to 4.15 pm due to the very high wind speed and temperature and very low relative humidity and pressure values as shown in Fig. 2.22 (a–d). A changing meteorological pattern is normal with high wind speed. The average daytime turbulence intensity is about $5.632 \cdot 10^{-14}$ m$^{-2/3}$. Further, minute fluctuations are obtained in the friction velocity, characteristic temperature and C_n^2 when the wind speed oscillates around a small value. Typically, the stronger minima are observed at both ends of the day. The larger fluctuations in the meteorological values from LT 6.30 am to 11.00 am generate unrealistically large sporadic C_n^2 values. Sustained and very smooth steady variations are seen in the nighttime.

2.8.5 Data for 16th November 2013, rainy

The weather parameters varied from 0.97 m s^{-1} to 4 m s^{-1} with SD of 1 m s^{-1} for wind speed, 21 °C to 25 °C with SD of 1 °C for temperature, 60 % to 100 % with SD of 10 % for relative humidity and 100.6 kPa to 101.1 kPa with SD of 0.1727 kPa for barometric pressure as shown in Fig. 2.23 (a–d). The different atmospheric conditions observed

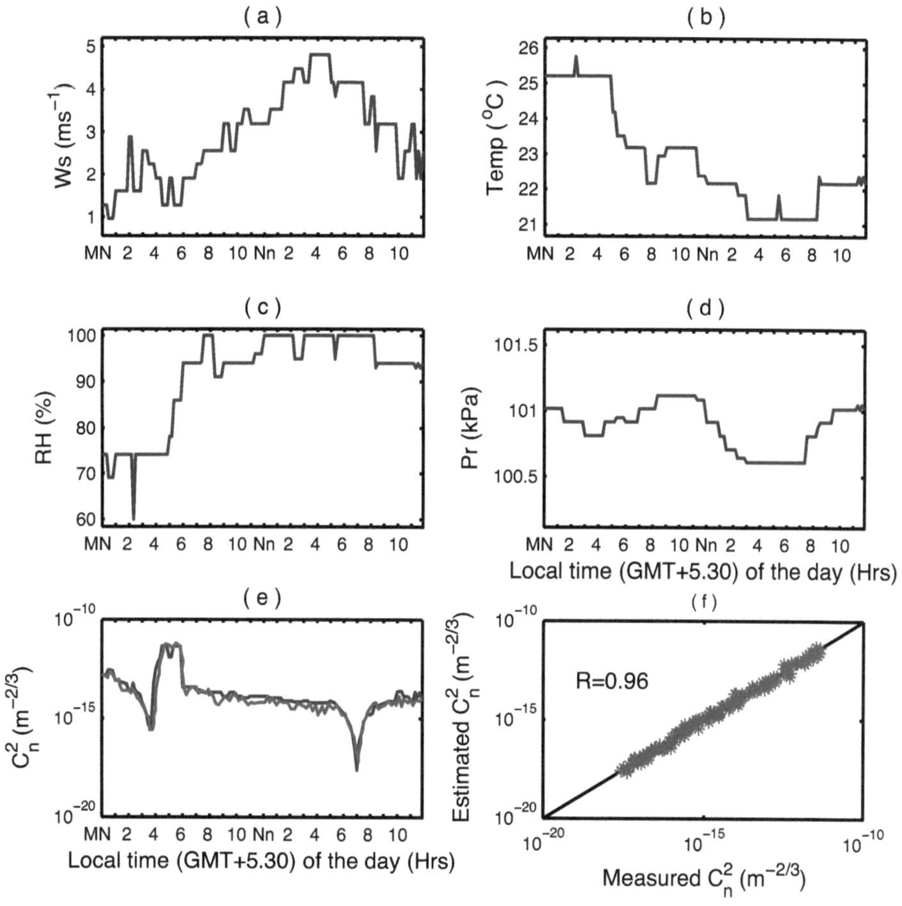

Fig. 2.23: Diurnal time series profile of meteorological parameters over a one day period. (a–d) A comparison of time series plots of atmospheric turbulence strength (C_n^2) estimated using PAMELA model (blue) and direct transmission measurement using scintillometer setup, i.e. based on signal intensity ensemble average data (red). (e) Correlation statistic plot for measured and estimated C_n^2 values. (f) MN: midnight and Nn: noon.

on 16/11/2013 (Saturday) are mostly cloudy, haze, light rain, light drizzle and heavy rain. The SD of Ws and RH are high while T and Pr are low. The experimental local time series plot of C_n^2 data corresponding to the local meteorological data collected on 16th November 2013 proved that the estimated values could yield a good correlation to the measured values at almost all times of day as shown in Fig. 2.23 (e).

The measurement analysis reveals that the correlation coefficient (R) flicks in the inner scale between 98 % and 99 % and the coefficient of determination (R^2) between 97 % and 98 %. Figure 2.23 (e) shows a very low turbulence intensity $\approx 2.326 \cdot 10^{-18}\ m^{-2/3}$ about 7.00 pm. A few short intervals of very high turbulence intensity $\approx 4.505 \cdot 10^{-12}\ m^{-2/3}$ are seen about 5.00 am due to low wind speed, high relative humidity, normal pressure, and relatively high temperature values as shown in Fig. 2.23 (a–d). The average daytime turbulence intensity is about $4.441 \cdot 10^{-15}\ m^{-2/3}$. Typically, the stronger minima are observed only in the evening. The smooth variations in the meteorological values generate almost flat behavior in the C_n^2 values. Sustained and smooth steady variations are seen in the daytime. The maximum value of $C_n^2 \approx 2.326 \cdot 10^{-18}\ m^{-2/3}$ is observed around 5.00 am as shown in Fig. 2.23 (e).

Thus the estimation of more accurate turbulence strength at our test field for different outdoor environmental conditions using the PAMELA model is difficult. Therefore, developing an empirical model based on the macroscale meteorological measurements to yield a more accurate estimate for local turbulence strength is important; and it is a matter of subsequent research.

2.9 Advantages

- Measuring ability and accuracy make the proposed system more viable and less costly for various applications related to environmental change.
- This system reduces trial experiments' time and cost.
- This method is completely generalized and problem independent and hence it can be easily rehashed to other parameter measurements.
- Any specific/separate interfacing hardware/data acquisition card, boards and software are not required.
- Moreover the cost of the existing standalone measurement instrument is high, further customization is additionally increasing the cost.
- Far more tasks could be performed along with the measurement modules.

2.10 Summary

The establishment of a scintillometer experimental setup for the link range of 0.5 km at an altitude of 15.25 m is described and the values of C_n^2 are measured. The meteorological sensors are connected with separate architectures created inside the FPGA to transfer the commands/address and readout the measurement data. A UART communication architecture as per the RS232 standard data frame format is also developed to transfer the measurement values to the data logging computer. A customized code is developed in the MATLAB environment to read the serial port data, manipulate them to find the true values, update the value and real-time plots and to perform statistical analysis. The measurements are continued for several days in a diurnal period cycle. The measurement performance of the proposed system is calibrated against the standard measurement instruments and an overall 98 % correlation accuracy is achieved. The expanded uncertainties of measurements are estimated and the variations are reported. C_n^2 is estimated using the PAMELA model with measured data. The estimated values of C_n^2 are validated with the scintillometer measurement values and the minimum and maximum values of correlation coefficient 0.79 and 0.97 respectively are observed from the investigation of five days' data collected in different seasons over a one year period. Moreover, a close correlation between estimated and measured C_n^2 is obtained for several days in clear outdoor environment conditions and correlation coefficients oscillate almost closely around unity, which confirms the proposed measurement system's accuracy. However, the maximum amount of deviation error of ± 0.000026 is observed on some days especially in high turbulence and calm regime due to deficiencies in the PAMELA model. These results trigger a way to develop a model with very accurate estimations at our test field to find the actual data rate the system is capable of operating under for different outdoor local environmental conditions (Bazil Raj et al. 2014a, with permission of IET).

3 Comparison of different models for ground-level atmospheric attenuation and turbulence strength (C_n^2) prediction with new models according to local weather data for FSO applications

Atmospheric parameters strongly affect the performance of Free Space Optical Communication (FSOC) systems when the optical wave is propagating through an inhomogeneous turbulence transmission medium. Developing models to get an accurate prediction of both optical attenuation and turbulence strength according to meteorological parameters becomes significant to understanding the behavior of a channel in different seasons and to apply suitable adaptive techniques to permit data transmission through the atmosphere. A dedicated Free Space Optical (FSO) link for the range of 0.5 km at an altitude of 15.25 m is established and explained. The power level and beam centroid information of the received signal are continuously and simoultaneously measured with the meteorological parameters using the optoelectronic assembly and the developed weather station respectively, and stored in a data logging computer. The existing models selected for comparative analysis in both the cases, based on exhibiting relatively less error, are described. Measured meteorological parameters (as input factors) and optical attenuation and turbulence strength (as response factor) of size [2000 × 4] are separately used for linear regression analysis and to design the mathematical models more suitable at the test field. Along with the model formulation methodologies, contribution of input factors' individual and combined effects on the response surface and coefficient of determination (R^2) estimated using analysis of variance (ANOVA) tools are presented. Cubic equation ($R^2 = 98.76\,\%$) and model equation V ($R^2 = 98.93\,\%$) are finalized for predicting optical attenuation and turbulence strength respectively. In addition, the prediction accuracy of the proposed and selected models for different seasons: monsoon, rainy, winter, presummer and summer in a one year period are investigated and validated in terms of Root Mean Square Error (RMSE) and Sum of Absolute Error (SAE). The average RMSE of 0.041 dB/km for optical attenuation and SAE of $0.000641 \cdot 10^{-9}\ m^{-2/3}$ for turbulence strength are achieved in the longer range dynamic of meteorological parameters in different local seasons.

3.1 Introduction

Free Space Optical Communication (FSOC) is a potentially high capacity and cost effective technique that is receiving growing attention and commercial interest (Nadeem & Leitgeb 2010; Fares & Adachi 2010; Das 2012; Awan et al. 2009c; Grabner & Kvicera 2011). Free Space Optics (FSO) is an age old technology that entails the transmission of information-laden optical radiation through the atmosphere from one point to other (Ghassemlooy & Popoola 2010). The limiting factor in FSOC is intensity noise due to fluctuations in atmospheric parameters, so called capacity limiting factors, which depend on the local meteorological conditions that cause the propagating laser beam to be deflected and/or scattered (Bouchet et al. 2006; Ni et al. 2009; Silva et al. 2011a; Zamek & Yitzhaky 2006; Rajbhandari et al. 2011; Tunick 2008). The scattering coefficient is defined as the ratio of original light intensity (scene irradiance) to attenuated scene intensity (irradiance) (Mudge et al. 2011; Recolons et al. 2007; Malik & Majumder 2013). While these deviations are small locally, the effects accumulate over the propagation path and can lead to scintillation, beam wandering and wavefront distortion that varies on timescales typically to the order of a millisecond or longer (Mudge et al. 2011; Leclerc et al. 2010; Sadot & Kopeika 1992; Doss-Hammel et al. 2004) which reduce the overall system reliability. Hence, it is reasonable that in a real FSOC environment, optical channels will appear to have randomly time varying characteristics which are difficult to predict/simulate (Chadha 2005; Giggenbach & Horwath 2005; Jurado et al. 2006b). Different weather conditions including dust, eddies of air of various size and velocity, fog and smoke are the causes that could potentially disrupt the FSOC by attenuating the input optical signal to the receiving side (Yitzhaky et al. 1997; Bendersky et al. 2004; Dror et al. 1996; Monin & Obukhov 1954). Optical attenuation varies from 0.2 dB/km to 480 dB/km in a very clear to moderate to high weather conditions (Ni et al. 2009; Sadot & Kopeika 1992; Bendersky et al. 2004; Monin & Obukhov 1954). Rain can cause attenuation up to 20–30 dB/km at a rain rate of 150 mm/h and snow can cause > 45 dB/km of loss (Awan et al. 2009b; Bouchet et al. 2006).

Fig. 3.1: Bird's eye view (GPS) of optical propagation path for atmospheric attenuation and C_n^2 field data measurement. The receiver system is located on the building at the left (marked by red balloon) and the transmitter system is located on the tower at the right (marked by blue balloon).

The disturbance of turbulence varies greatly with location and time and its strength is usually greater near the ground and falls-off exponentially with increasing altitude due to wind shear. As a result of the exponential distribution, the effects of turbulence are much less severing in a near-vertical path than in a horizontal propagation path of comparable length (Monin & Obukhov 1954). From this point of view, real-time experimental studies on the atmospheric attenuation and turbulence strength (C_n^2) (Chang et al. 2006) as a function of meteorological parameters are important to understand the channel effects on the propagating optical beam to characterize the FSO channel that the maximum data rate system could operate at (Awan et al. 2009c; Grabner & Kvicera 2011) at the test field and this is the main contribution in this paper. The Earth Observation Satellite (EOS) view of the experimentation field location representing the transmitter and receiver established for these studies is shown in Fig. 3.1. The AC plant, mechanical workshop, civil engineering block, playground and agricultural land are the terrain disturbance sources in addition to the atmospheric changes to the optical pathway. The transmitter lab is located on a tower (lat. 10°38′46.7334″ and long. 79°3′12.0774″) while the receiver is built on the rooftop of the existing information technology (IT) block (lat. 10°38′52.8468″ and long. 79°2′56.6268″) as in Fig. 3.1. Meteorological sensors are deployed close to transmitter, receiver and at the midway of the optical link (Sofiya et al. 2015). The wireless data logging hardware circuits are connected with the receiver unit and the measured weather data are recorded in the computer.

The fundamental law to estimate attenuation of an optical signal on the basis of the atmospheric visibility (V) in km is the Koschmieder law, which defines the visibility as the distance to an object/optical intensity at which the visual contrast/transmittance of an optical signal drops to a certain value of the visual/transmittance threshold T_{th} of the original visual contrast/intensity (100 %) along the propagation path (Grabner & Kvicera 2007; Gebhart et al. 2004; Ijaz et al. 2013b; Prokes 2009). The meteorological visibility can be therefore expressed with the Beer–Lambert law as (Ijaz et al. 2013a; Malm 1999)

$$V = \frac{10 \log_{10}(T_{th})}{\beta_\lambda} = \frac{4.343 \log_{10}(T_{th}) R}{\log_{10}(T)}, \tag{3.1}$$

where T_{th} is the transmission threshold in %, β_λ is the atmospheric attenuation coefficient in dB/km, R is the link range in km and T is the transmittance in dB as given by in Bouchet et al. (2006). In Eq. (3.1), the smaller value of T_{th} defines a larger visibility range and in 1924 Koschmieder defined a value of T_{th} as 2 % (5 % for aeronautical requirements) following Helmholtz theory. The selection of T_{th} varies the definition of visibility and the corresponding attenuation of the optical signal for a given visibility range (Grabner & Kvicera 2008). The transmittance (total loss) is measured using an appropriate power meter as received power (P_R)/transmitted power (P_T), both in mW at the peak wavelength of 550 nm in visible spectrum. The transmittance of the 850 nm is also simultaneously measured. Using the optical signal or video/still camera image or transfer function (Malik & Majumder 2013) are the most common methods for attenuation measurement. The optical beam attenuation along the propagation path is directly measured using an optical signal in dB/km as

$$\text{Attenuation} = \left(\frac{10\log_{10}(\mathrm{P_T}/\mathrm{P_R})}{\mathrm{R}}\right). \tag{3.2}$$

Since the atmospheric particle concentration and size distribution vary in the spatial and time domains, it becomes difficult to predict the atmospheric attenuation using the theoretical approach (Bouchet et al. 2006). Therefore, predictions of atmospheric attenuation and turbulence strength using empirical models are increasing. Most of the attenuation empirical models use visibility data alone in order to characterize attenuation and concentrate only on very low visibility conditions which never or very rarely occur at our test field. Although various models including Kruse, Kim, Naboulsi, Ferdinandov, Fischer, Grabner, et al.; Pierce et al.; Pesek et al; Ijaz et al.; and Nadeem et al. are available to estimate atmospheric attenuation, no model provides generalizations. The sufficient criticism of the existing models and the root-cause for/of the new models can be found in several recent publications including (Nadeem & Leitgeb 2010; Grabner & Kvicera 2011, 2007; Zamek & Yitzhaky 2006; Ijaz et al. 2013a; Randall 2004).

Most of these models were developed for single/limited wavelength (Nadeem & Leitgeb 2010; Grabner & Kvicera 2011, 2007; Ijaz et al. 2013b), using a short range of optical link (Grabner & Kvicera 2011, 2007), an indoor optical link (Rajbhandari et al. 2011), a manmade controlled atmospheric chamber with smaller data sets (Ijaz et al. 2013b), and Liquid Water Content (LWC) (Khan et al. 2012). These models are explicitly experimentally investigated to test the robustness at our test field. The experimental outcomes clearly demonstrate that in spite of the significant number of investigations the prediction results of these models contradict each other when applied in the real-world open atmosphere and exhibit less correlation with the measurement data acquired for several days in different seasons. Development of an accurate and locally valid model to predict the atmospheric attenuation and turbulence strength according to local weather data becomes interesting and important (Elkamchouchi & El-Shimy 2006) and the main contribution of this chapter is to design a suitable mitigation technique to improve the overall performance of the FSOC system.

The rest of this chapter is organized as follows: Section 3.2 presents the background and related works, Section 3.3 describes the transmitter and receiver experimental setups and optoelectronic assembly, Section 3.4 reviews some of the existing models selected for the comparative analysis, Section 3.5 demonstrates the development and validation of the proposed models of atmospheric attenuation and turbulence strength, Section 3.6 discusses the experimental results and data analysis, and Section 3.7 draws some conclusions.

3.2 Background and related works

In the last few years, a lot of in situ field measurements related to the modeling of atmospheric attenuation and turbulence strength (C_n^2) have been carried out and can be found in the literature.

Nadeem & Leitgeb (2010) discuss Krus, Kim Al Naboulsi advection and convection models. A model is proposed in terms of visibility to estimate atmospheric attenuation. The results of all these models are compared and the closeness with the measured values is analyzed in terms of Sum Squared Error (SSE) and Root Mean Square Error (RMSE).

Awan et al. (2009c) discuss the geographical location of the experimental setup. The basics of fog microphysics, density measurement and relations between relative humidity and temperature with optical attenuation are explained. Optical attenuation in fog and snow conditions is detailed. Finally, link budget and the experimental data are analyzed.

Ni et al. (2009) developed an RF-FSO test link and studied atmospheric turbulence effects on these two links. Turbulence strength is estimated using the Rytov variance and Angle-of-Arrival (AoA) methods for the link range of 300 m between two buildings in Hamamatsu Photonics K.K.

Vinicius N. H. Silva et al. (2011b) used the beam wandering based technique (triangulation-like) to estimate the C_n^2 using 21 cm optical link distance. It is similar to the optoelectronic position detector (OPD) where the radial displacement is the principal component. The optical beam tracking and simulation results for various displacements are discussed.

Arnold Tunick (2008) discusses the geographical and terrain location of the experimental setup. The method of environmental and C_n^2 data acquisition and analysis are given. The data correlation and curve fit equations are discussed. The influence of wind vector is discussed.

Jaume Recolons et al. (2007) discuss various theoretical models on RMS of centroid displacement, beam wander variance and spot size, hot spot displacement, beam parameters, profile of untracked and tracked beam, beam radius, mean irradiance profile and scintillation index. The simulation results are discussed.

Troy T. Leclerc et al. (2010) describe the optical and meteorological equipment used in their experimental setup. Macro-meteorological models (as a function of weather data) developed/modified by Bendersky Kopeika and Blaunstein (BKB model) are given. The BKB model utilized the $-4/3$ height scaling power law introduced by Walters and Kunkel and the experimental data corresponding to various days are presented. The optical path distance of 65 m is used.

Sadot and Kopeika (1992) describe various methods of C_n^2 prediction. The experimental setup and location (Tx and Rx) information are detailed. A regression model is presented and experimental data are analyzed. The predicted values are validated against the measurement data.

Yitzhak Yitzhaky et al. (1997) present a macro-meteorological model for predicting C_n^2 and aerosol modulation transfer function. The restorations of atmospherically blurred images are the principal components. Image restoration and C_n^2 estimation results are reported.

Sergey Bendersky et al. (2004) clearly explain the C_n^2 experimental site, instrumentation and measurement data. A detailed analysis of Monin–Obuhkov similarity models are given. The statistical data corresponding to the deviations among these models and against the measurement data are explained.

Arun K. Majumdar et al. (2006) discuss the theoretical concepts of atmospheric turbulence effects. Various atmospheric turbulence models (as a function of altitude) for estimating turbulence strength, coherence strength, isoplantic angle, Rytov variance and Greenwood frequency are explained. Experimental procedure and data collected on different days are discussed.

Micheal Gebhart et al. (2004) detail the receiver calibration measurement with sample computation examples. The location of the measurement setup is described. The experimental results of visibility and optical attenuations are given. The estimation results are discussed.

Ales Prokes (2009) discusses the weather dependency and performance measures of the FSOC system. The estimation methods and formulas of link budget, propagation loss, power link margin, scattering attenuation and turbulence power loss and link availability are described in detail. Simulation results are presented and the data are analyzed.

Muhammad Ijaz et al. (2013b) discuss the theory related to visibility and attenuation. The construction of the controlled laboratory atmospheric chamber is explained. The Krus model is slightly modified and validated for different wavelengths with the attenuation data acquired using the indoor atmospheric chamber.

Usman et al. (2013) propose models to estimate visibility as a function of meteorological data (atmospheric temperature and relative humidity). The fraction of sunshine hours is used in modeling. Measured, predicted visibility and Percentage of Absolute Error (PAE) are analyzed.

Arnold Tunick (2007b) describes the optoelectronics experimental setup configuration for optical turbulence measurements over a 2.33 km free-space laser path. The Rytov variance approach is followed to measure the C_n^2 and the values are compared against the scintillometer data.

Steve Zamek and Yitzhak Yitzhaky (2006) describe the methods of estimating angle of arrival variance and spatial and temporal correlation properties. C_n^2 is estimated from the blurred image and the results are presented.

3.3 Experimental setup and description of optoelectronic assembly

Transmitter and receiver laboratories are established for the link range of 0.5 km at an altitude of 15.25 m exclusively for studies on the prediction of atmospheric attenuation and turbulence strength according to local weather data. A semiconductor laser source, driver and power supply unit and transmitting optics are the optoelectronics devices at the transmitter. The transmitting optics is used to expand the incident beam of diameter 3 mm to 9 mm to reduce beam divergence at the aperture of the telescope. The receiver laboratory consists of telescope, Narrow Band Interference Optical Filter (NBIOF), variable beam splitter, photodiode, Optoelectronics Position Detector (OPD), Mono-Pulse Arithmetic Circuit (MPAC), power meter, automated data acquisition system and data logging in Personal Computer (PC). The optoelectronics equipment is assembled so as to measure power fluctuation in mW and beam displacement in V as suggested by Gupta et al. (2006), Andrews et al. (2001) and Vladutescu et al. (2012).

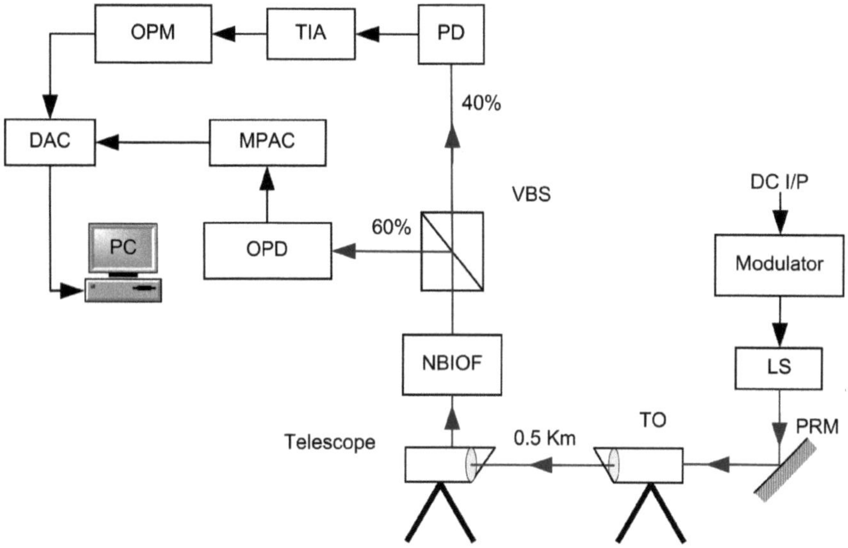

Fig. 3.2: Schematic diagram of Laser Communication Laboratory (LCL) experimental setup (transmitter and receiver) constructed to measure atmospheric attenuation and turbulence strength and modeling.

The telescope captures all the photons and concentrates them to the NBIOF which is used to permit only the 850 nm (center wavelength) optical beam as well as to block the other wavelengths. The variable beam splitter divides the incident beam (output of the NBIOF) into two directions: reflecting and propagating beam. The reflected beam is made to fall on the OPD to measure the beam wandering information using the MPAC computations. The propagating beam is made to fall on the photodiode and its output is connected to the power meter. The schematic diagram of the experimental setup is shown in Fig. 3.2 and all the optoelectronic devices are mounted on vibration damped optical breadboards as shown in Fig. 3.3. A low cost and high accuracy integrated weather station is built with specialized sensors to continuously measure the wind speed (Ws), temperature (T), relative humidity (RH), and pressure (P).

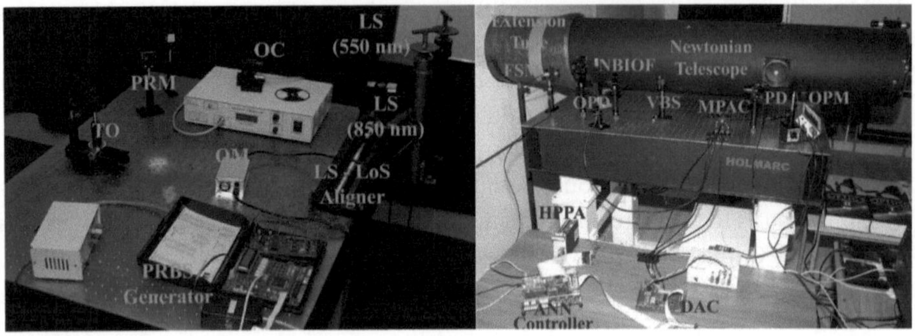

Fig. 3.3: Snapshot showing the optoelectronic components assembly on the vibration damped optical breadboard (a) transmitter and (b) receiver.

The weather station is deployed at an altitude of 15.25 m near to the transmitter, receiver and at the midpoint (above civil block). The meteorological data were continuously acquired every second during different outdoor environmental conditions (local seasons) and the daywise data are averaged over 5 minute periods. A separate Graphical User Interface (GUI) is developed using the MATLAB environment (Smith 2006) for automated data acquisition, data logging and on/off line computations. The main optoelectronics devices and their parameters are given in Tab. 3.1.

The necessary parameters are extracted from the measurement values and applied to the models to predict the attenuation in dB/km and turbulence strength in $m^{-2/3}$.

3.4 Comparison models of atmospheric influence on optical propagation

The optical radiation traveling through the atmospheric turbulent channel interacts with the molecular constituents of the atmosphere and causes some of the photons to extinguish. This event ultimately results in power loss, temporal and spatial distortion that strongly depends on the local weather conditions/seasons (Bouchet et al. 2006). Although various wavelength independent/dependent models are available, the models exhibiting relatively less RMS are considered for the comparative analysis with the

Tab. 3.1: Experiment parameters of the optical link.

Parameter		Value
Transmitter		
Laser diode	Peak wavelength	850 nm
	Maximum optical power	10 mW
	Beam size at aperture	3 mm
	Beam divergence	3 mrad
	Laser beam propagation model	Plane
Optical Lens	Diameter	3.9 mm
Channel	Range	0.5 km
	Altitude	15.25 m
	Surface roughness length	0.03 m
Receiver		
Telescope	Aperture	330.9 mm
	Type	Newtonian
Filter	Type	NBOIF
	CWL	850 nm
Optical collimator	Collimation ratio	9 : 3
Beam splitter	Type	Variable
	Splitting range	0–100 %
OPD	Error output	±10 V
	Position sensitivity	±2 mm
	Error computation	MPA
Photodetector	Active area	1 mm^2
	Half angle field of view	±75°
	Spectral sensitivity	0.59 A/W
	Rise and fall time	5 ns
Power meter	Thorlabs PM100D	400–1100 nm

new models. The selected models are reviewed in this section with their background and limitations. The detailed descriptions of the selected models can be found in the references.

3.4.1 Atmospheric attenuation

Atmospheric attenuation results from a combined effect of absorption and dispersion of the optical field by the turbidity suspended in the medium dependent on the local atmospheric and terrain conditions. The molecular and aerosol behavior for the scattering and absorption process is wavelength dependent. The atmospheric windows showing the transmittance of the optical wireless signal can be found in Bouchet et al. (2006) and Kim et al. (2001). Molecular scattering is very small in the near infrared range due to dependency on λ^{-4} and thus can be neglected. Therefore, aerosol scattering becomes the dominating factor along with the variations of meteorological parameters reducing the total extinction coefficient. The behavior of an atmospheric attenuation model has to be explicitly and experimentally verified for its generalization and prediction accuracy when applied in real-world open atmosphere for the most commonly using FSOC wavelength of 850 nm. The models considered for the comprehensive analysis are discussed below and the attenuation is estimated in dB/km (Gebhart et al. 2004).

3.4.1.1 M. Ijaz's model

Muhammad Ijaz and coworkers have recently developed an empirical model using the data set acquired from the indoor manmade controlled optical turbulence chamber in order to predict the specific attenuation of fog and smoke for different wavelength as (Ijaz et al. 2013b)

$$\beta_\lambda = \frac{17}{V}\left(\frac{\lambda}{\lambda_0 = 550\,\text{nm}}\right)^{-q(t)}, \tag{3.3}$$

where V is visibility in km, λ is source wavelength in nm, $q(t) = 0.1428\lambda - 0.0947$ for fog and $0.8467\lambda - 0.5212$ for smoke. The general expression for q(t) is obtained using empirical curve fitting of some experimental data with a reference wavelength of 550 nm. The experiment is conducted in a controlled environment where the fog and smoke are individually controlled which would not be possible in real FSO since the channel is uncontrolled. Further, the atmosphere is not only comprised of fog and/or smoke, but it also consists of different parameters based on the terrain, geographical and local climate. However, this model is valid only for $550\,\text{nm} < \lambda < 1600\,\text{nm}$ and $0.015\,\text{km} < V < 1\,\text{km}$ (Ijaz et al. 2013b). The frequency of this visibility at our test field is much lower. This model is developed from empirical data. It is observed that the intervention of the optical link by the smoke at our test field is also much lower, therefore, the smoke function is neglected from our comparative analysis.

3.4.1.2 M. S. Awan's model

Muhammad Saleem Awan and coworkers have proposed two models for predicting the attenuation coefficient based on Gaussian fitting by employing the nonlinear least squares method and using the trust-region reflective Newton algorithm (Awan et al. 2009c). The variables of the first and second models are relative humidity in % and temperature in °C respectively.

$$\beta_{\text{atten}} = 25.75 \exp\left(-\left(\frac{\text{RH} - 83.85}{1.026}\right)^2\right) + 93.89 \exp\left(-\left(\frac{\text{RH} - 113.8}{21.77}\right)^2\right) + 24.46 \exp\left(-\left(\frac{\text{RH} - 85.64}{0.4174}\right)^2\right) \quad (3.4)$$

$$\beta_{\text{atten}} = 36.04 \exp\left(-\left(\frac{\text{T} - 3.839}{0.1298}\right)^2\right) + 26.16 \exp\left(-\left(\frac{\text{T} - 4.189}{0.1722}\right)^2\right) + 263.7 \exp\left(-\left(\frac{\text{T} + 12.74}{9.071}\right)^2\right) \quad (3.5)$$

The RMSE value of attenuation predicted using temperature is relatively high, therefore this model is not considered for verification and relative humidity based model, Eq. (3.4) alone is considered.

3.4.1.3 Itai Dror's model

Itai Dror and coworkers have developed a model using regressive analysis and Statistical Analysis System (SAS) software as a function of meteorological parameters to predict the atmospheric extinction coefficient as (Dror et al. 1996)

$$\sigma_{\text{ext}} = \frac{1}{(1 - \text{RH}/100)^4} \exp\left(-\left(a_0 + \sum_{i=1}^{5} \beta_i X_i\right)\right) \quad (3.6)$$

where a_0 is a constant equal to 10.01, β_i are regression coefficients and X_i are meteorological parameters. The values of β_i are: $\beta_1 = -13.25$, $\beta_2 = 5.64$, $\beta_3 = 0.0007678$, $\beta_4 = -0.0000281$, and $\beta_5 = 0.002408$ and X_i are: $X_1 = 1 - \text{RH}/100$, $X_2 = (1 - \text{RH}/100)^2$, $X_3 = \text{T}/(1-\text{RH}/100)^3$, $X_4 = \text{T}^2/(1-\text{RH}/100)^3$, and $X_5 = (1-\text{RH}/100)^3\text{Ws}^3$. The regression coefficients for the wavelength of 850 nm are not directly available and obtained by the curve fitting technique. The inputs dynamic limitation ranges of the model are wind speed: 0 to 8 m s^{-1}, temperature: 15 to 45 °C and relative humidity: 10 to 85 % (Dror et al. 1996). It is observed that the variation of the meteorological parameters exceeds these dynamic ranges at our test field.

3.4.1.4 Bataille's model

Bataille has proposed a model to predict specific extinction coefficient (σ_m) in the lower atmosphere in terms of atmospheric temperature and relative humidity as (Bouchet et al. 2006; Bataille 1992)

$$\sigma_m = -\ln\left(B_1 + B_2\text{T}' + B_3\text{RH} + B_4\text{T}'\text{RH} + B_5\text{T}'^2 + B_6\text{RH}^2 + B_7\text{T}'\text{RH}^2 + B_8\text{T}'^2\text{RH} + B_9\text{RH}^3 + B_{10}\text{T}'^3\right) \quad (3.7)$$

where $\text{T}' = \text{T}/273.15$ is the reduced temperature of the air and RH is the absolute humidity in g m^{-3}. The coefficients are $B_1 = 0.9953$, $B_2 = 0.01311$, $B_3 = -0.00148$, $B_4 =$

0.00137, B_5 = −0.012162, B_6 = 0.00000038, B_7 = 0.00000026, B_8 = −0.0004221, B_9 = 0, and B_{10} = 0.003763. This is an inverse model and derived from empirical data.

3.4.2 Atmospheric optical turbulence strength

Most of the models found in the literature predict the C_n^2 as a function of altitude and their prediction results greatly differ from measurement data of horizontal path turbulence. Furthermore, some of the models have been developed for specific locations like rocky terrain, Hufnagel–Valley, Submarine field, Greenwood and Gurvich, etc. C_n^2 not only varies as a function of altitude, but also according to local meteorological conditions, geographic location, terrain type and time of day (Doss-Hammel et al. 2004; Majumdar et al. 2006; Font et al. 2006; Yague et al. 2001). Therefore, only models predicting horizontal turbulence strength and exhibiting low RMSE are considered for the comparative analysis.

3.4.2.1 PAMELA model

According to the PAMELA model, turbulence fluctuation in the surface boundary layer is a function of altitude (h), local conditions (terrain type), geographical location, cloud cover, meteorological values, latitude, longitude, number of days in year, Greenwich Mean Time (GMT), surface roughness length, and local time of day (Doss-Hammel et al. 2004). The PAMELA model given in Eq. (3.8) provides a C_n^2 estimation within the surface boundary layer and it accepts all the above parameters of the test field as the inputs. The meteorological parameters measured are entered into the model to estimate turbulence strength.

$$C_n^2 = 5.152\phi_h \left(\frac{1}{\phi_m - \zeta}\right)^{0.33} \left(\frac{77.6 \cdot 10^{-6} P}{T^2}\right)^2 h^{-0.667} \left(\frac{-H}{C_p \rho u_*}\right)^2 , \tag{3.8}$$

where ϕ_h is temperature gradient, ϕ_m is wind shear estimated from the wind speed, ζ is eddy dissipation rate, H is heat flux, C_p is specific heat, ρ is mass density and u_* is friction velocity. As can be seen from Eq. (3.8), $C_n^2 \to \infty$ as Ws $\to$ 0 (Doss-Hammel et al. 2004; Font et al. 2006), therefore, the minimum wind speed is bounded away from zero i.e. 0.27 m s^{-1}.

3.4.2.2 Hufnagel–Valley model

The Hufnagel–Valley model predicts the C_n^2 for inland sites and daytime viewing conditions since it permits variations in altitude, wind speed and near ground turbulence levels. In this model, a sum of three exponential decay terms corresponding to a surface boundary layer, a strong layer caused by the high altitude jet stream and a background tropopause layer are present as in (Majumdar et al. 2006)

$$C_n^2(h) = A\exp(-h/100) + 5.94 \cdot 10^{-53}(v/27)^2 h^{10}\exp(-h/1000) + 2.7 \cdot 10^{-16}\exp(-h/1500)\,, \tag{3.9}$$

where A is the nominal value of C_n^2 in $m^{-2/3}$ at the ground and v is the Root Mean Square (RMS) wind speed in $m\,s^{-1}$.

3.4.2.3 Beam wandering model

The beam wander effect is related to the displacement of the instantaneous center of the beam: the point of maximum irradiance of a traveling wave over the receiver plane. It is well known that this phenomenon is caused by the large-scale inhomogeneities due to their refractive effects (Silva et al. 2011b; Recolons et al. 2007). In the 2D Cartesian or polar coordinate (x, y or r, θ) receiver plane, the stochastic polar variable γ_c (radial displacement) of the beam center is computed by coordination transformation as (Silva et al. 2011b)

$$\gamma_c = \sqrt{x_c^2 + y_c^2}\,, \tag{3.10}$$

where x_c and y_c are the Cartesian coordinates of the center of the Gaussian light spot on OPD and used to track the beam wander with respect to origin (0, 0) mm. Beam wander can be statistically characterized by the variance of γ_c and related to atmospheric turbulence strength using the geometrical optics approximation as (Das 2012; Silva et al. 2011b; Tunick 2008)

$$C_n^2 = \frac{\langle \gamma_c^2 \rangle}{2.42 R^3 W^{-1/3}}\,, \tag{3.11}$$

where W is beam waist (width) of the Gaussian beam in mm and R is length of the optical link in km.

3.4.2.4 Polynomial regression

This model is developed to obtain the best estimation of C_n^2 according to the macroscale meteorological data in situ. The concept of temporal or solar hour is introduced. The temporal hour at sunrise is 00:00, at noon it is 06:00 and at sunset it is 12:00 in any day. Further, it is allowed to have negative values. The polynomial regression model is (Leclerc et al. 2010; Bendersky et al. 2004)

$$C_n^2 = 3.8 \cdot 10^{-14}W + 2 \cdot 10^{-15}T - 2.8 \cdot 10^{-15}RH + 2.9 \cdot 10^{-17}RH^2 - 1.1 \cdot 10^{-19}RH^3 - 2.5 \cdot 10^{-15}Ws + 1.2 \cdot 10^{-15}Ws^2 - 8.5 \cdot 10^{-17}Ws^3 - 5.3 \cdot 10^{-13}\,, \tag{3.12}$$

where W is temporal hour weight values taken from Leclerc et al. (2010) for computations. This model is the best one, especially in a practical manner since it requires only macroscale meteorological parameters which can be measured directly by a suitable weather station. This model is rehashed with the introduction of solar radiation (solar flux provides information concerning cloudiness) and aerosol loading in the atmosphere as (Bendersky et al. 2004)

$$\begin{aligned} C_n^2 &= 5.9 \cdot 10^{-15} \mathrm{W} + 1.6 \cdot 10^{-15} \mathrm{T} - 3.7 \cdot 10^{-15} \mathrm{RH} + 6.7 \cdot 10^{-17} \mathrm{RH}'^2 \\ &\quad - 3.9 \cdot 10^{-19} \mathrm{RH}^3 - 3.7 \cdot 10^{-15} \mathrm{Ws} + 1.3 \cdot 10^{-15} \mathrm{Ws}^2 - 8.2 \cdot 10^{-17} \mathrm{Ws}^3 \\ &\quad + 2.8 \cdot 10^{-14} \mathrm{SF} - 1.8 \cdot 10^{-14} \mathrm{TCSA} + 1.4 \cdot 10^{-14} \mathrm{TCSA}^2 - 3.9 \cdot 10^{-13} , \end{aligned} \tag{3.13}$$

where SF is solar flex ($\mathrm{kW\,m^{-2}}$) and TCSA is Total Cross Section Area ($\mathrm{cm^2/m^3}$). The TCSA can be determined by

$$\begin{aligned} \mathrm{TCSA} &= 9.96 \cdot 10^{-4} \mathrm{RH} - 2.75 \cdot 10^{-5} \mathrm{RH}^2 + 4.86 \cdot 10^{-7} \mathrm{RH}^3 \\ &\quad - 4.48 \cdot 10^{-9} \mathrm{RH}^4 + 1.66 \cdot 10^{-11} \mathrm{RH}^5 - 6.26 \cdot 10^{-3} \ln \mathrm{RH} \\ &\quad - 1.37 \cdot 10^{-5} \mathrm{SF}^4 + 7.30 \cdot 10^{-3} . \end{aligned} \tag{3.14}$$

The physical interpretation of this model can be found in Sadot & Kopeika (1992), Yitzhaky et al. (1997), and Bendersky et al. (2004). The limitations of this model are wind speed: 0 to 10 $\mathrm{m\,s^{-1}}$, temperature: 9 to 35 °C and relative humidity: 14 to 92 % (Das 2012).

3.5 Formulation of the mathematical model

This model provides a basis for making predictions about the outcomes of experiments and/or measurements. Direct methods for practical atmospheric problems are usually thwarted by the sheer size and complexity of the atmosphere. Most of the existing models, as pointed out by several authors, are derived from the data corresponding to their local atmospheric conditions; therefore, they failed to attain generalization on predicting optical attenuation and atmospheric turbulence strength (Nadeem & Leitgeb 2010; Recolons et al. 2007; Leclerc et al. 2010; Sadot & Kopeika 1992; Dror et al. 1996). Furthermore, these models do not offer any suitable means to tune (adapt) their parameters to fit a new test field (Awan et al. 2009c; Randall 2004). Therefore, having a more accurate prediction of local atmospheric attenuation and turbulence strength at the place of experimentation becomes significant which leads to proposing the new models. The formulations of the proposed models are detailed below:

3.5.1 Atmospheric attenuation

Pilot survey experiments were conducted using the developed experimental setup shown in Fig. 3.2 in different outdoor atmospheric conditions. The ensemble average of measured weather, visibility, transmittance, scintillation and attenuation data for the diurnal pattern are stored in the computer almost for a one year period. Independent historical formulation data set of size [2000 × 4], i.e. three input factors: wind speed (Ws) in $\mathrm{m\,s^{-1}}$, temperature (T) in °C and relative humidity (RH) in % and one response factor: optical attenuation (Atten) in dB/km, are given as the input to the Minitab software (Oehlert 2000). The optimistic regression models of different combinations of the input factors and orders at the confidence level of 98 % are formulated. The regression models developed for the prediction of optical attenuation according to weather data are as follows:

3.5.1.1 Linear equation

$$\text{Atten} = 12.4487 - 0.218101\text{Ws} - 0.0817053\text{T} - 0.115195\text{RH}\,. \tag{3.15}$$

3.5.1.2 Quadratic equation

$$\begin{aligned}\text{Atten} = {} & 10.3978 + 0.16793\text{Ws} - 0.225069\text{T} + 0.0570979\text{RH} \\ & - 0.0587006\text{Ws}^2 + 0.00202767\text{T}^2 - 0.00159008\text{RH}^2\end{aligned} \tag{3.16}$$

3.5.1.3 Interaction equation

$$\begin{aligned}\text{Atten} = {} & 28.6487 - 2.83927\text{Ws} - 0.373622\text{T} - 0.296224\text{RH} \\ & + 0.0364545\text{Ws}\,\text{T} + 0.023798\text{Ws}\,\text{RH} + 0.00242482\text{T}\,\text{RH}\end{aligned} \tag{3.17}$$

3.5.1.4 Full model equation

$$\begin{aligned}\text{Atten} = {} & 66.2563 - 0.932582\text{Ws} - 2.54336\text{T} - 0.609565\text{RH} \\ & + 0.0678237\text{Ws}^2 - 0.0297737\text{Ws}\,\text{T} + 0.0169019\text{Ws}\,\text{RH} \\ & + 0.0316436\text{T}^2 + 0.0113863\text{T}\,\text{RH} + 0.000653698\text{RH}^2\end{aligned} \tag{3.18}$$

3.5.1.5 Cubic equation

$$\begin{aligned}\text{Atten} = {} & -1.09734 - 3.18118\text{Ws} + 2.49896\text{T} + 0.277543\text{RH} \\ & + 0.0667605\text{Ws}^2 + 0.0381112\text{Ws}\,\text{T} + 0.0264432\text{Ws}\,\text{RH} \\ & - 0.106594\text{T}^2 + 0.00641653\text{T}\,\text{RH} - 0.0148677\text{RH}^2 \\ & - 0.0060649\text{Ws}^3 + 0.00118457\text{T}^3 + 9.56075\cdot 10^{-5}\text{RH}^3\end{aligned} \tag{3.19}$$

The individual and combined effects of the input parameters on the performance measures (prediction accuracy) are analyzed and subsequently mathematical models are formulated as in Eqs. (3.15)–(3.19). Further, practical constraints during the optimization process of the developed mathematical model are checked in parallel. The sensitivity of the atmospheric parameters to optical turbulence has been analyzed. Minitab

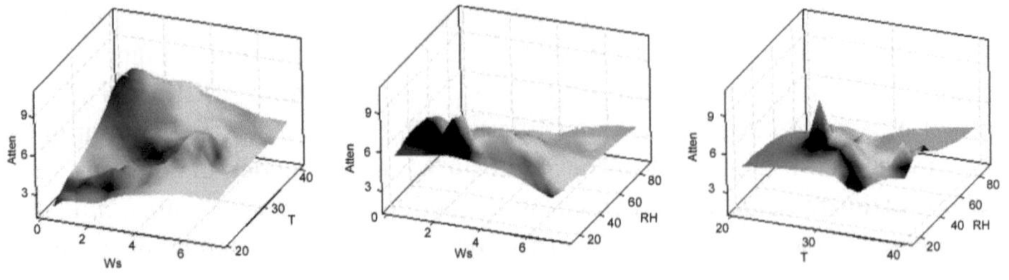

(a) (b) (c)

Fig. 3.4: Illustrations of input-output response surface plots. (a) attenuation vs wind speed and temperature, (b) attenuation vs wind speed and relative humidity, and (c) attenuation vs temperature and relative humidity

software's analysis of variance (ANOVA) tool is used to analyze the individual and combined effects of atmospheric parameters on optical attenuation. Figure 3.4 shows different possible combined effects of wind speed, temperature and relative humidity on optical attenuation.

The computed ANOVA results are used to understand the most dominating input factors on the response surface and those combinations are only considered while optimizing the model. The combinations that contribute less are ignored. Utmost prediction accuracy is considered while ignoring the less contributing combinations. Many models (Eqs. (3.15)–(3.19)) were examined in order to find a model more suitable for predicting atmospheric attenuation as a function of meteorological parameters. The coefficient of determination (R^2) values are computed using all the models, Eqs. (3.15)–(3.19), with new data sets, and the results are given in Tab. 3.2.

Tab. 3.2: R^2 value of Response Surface Model (RSM) for attenuation and turbulence strength prediction.

Regression Model	**R^2 value**	
	Attenuation (%)	**C_n^2 (%)**
Linear	79.04	NC
Linear + square	89.23	NC
Linear + interaction	90.37	NC
Linear + interaction + square	91.35	NC
Linear + square + interaction + cubic	98.76	NC
Model equation I	–	78.57
Model equation II	–	82.67
Model equation III	–	89.32
Model equation IV	–	92.48
Model equation V	–	98.93

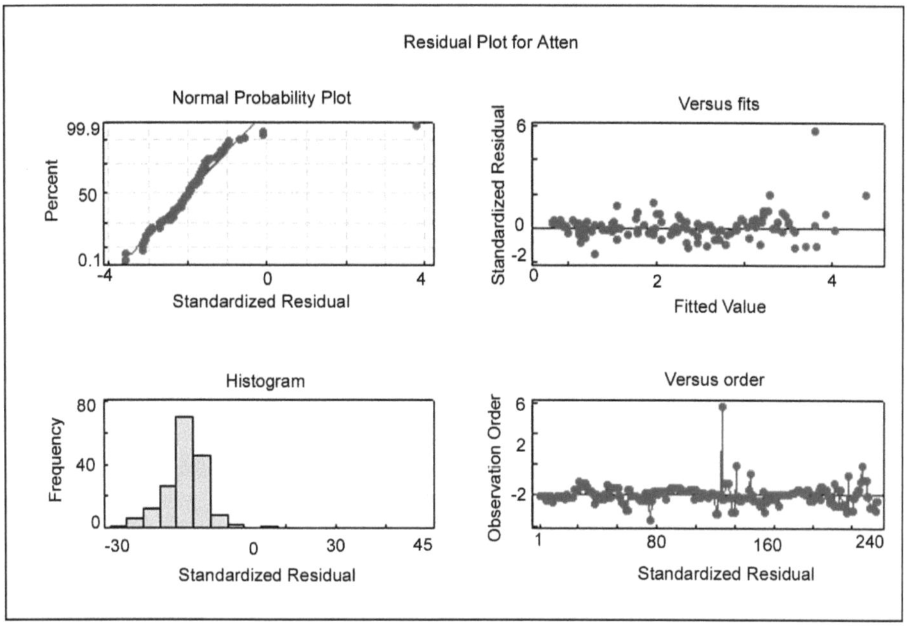

Fig. 3.5: Residual plots of the developed regression model (cubic equation) for optical attenuation.

The cubic model exhibits the greatest correlation with the measurement values, therefore, in the model Eq. (3.19) alone is considered for practical implementation for predicting optical attenuation according to local weather data. The residual plots obtained with the model Eq. (3.19) for optical attenuation are shown in Fig. 3.5.

3.5.2 Atmospheric turbulence strength (C_n^2) from meteorological measurements

Turbulence strength (C_n^2) values measured in different meteorological conditions are related to the meteorological parameters measured during the same time period over a particular day using the Minitab software and regression analyses are performed. The data sets of size [2000 × 4] i.e. three input factors (Ws, T, and RH) and one response factor (C_n^2) are used to formulate the regression model as given below:

3.5.2.1 Model equation I

$$\begin{aligned} C_n^2 = 1\cdot 10^{-14}(&399.774 + 4.88372\text{Ws} - 14.7804\text{T} - 5.58958\text{RH} \\ &- 2.32469\text{Ws}^2 + 0.203989\text{Ws}\,\text{T} - 0.0616557\text{Ws}\,\text{RH} \\ &+ 0.153508\text{T}\,\text{RH} + 0.170863\text{Ws}^3 + 0.00276336\text{T}^3 \\ &+ 0.000116384\text{T}\,\text{RH}^2 + 9.30618\cdot 10^{-5}\text{RH}^3) \end{aligned} \tag{3.20}$$

3.5.2.2 Model equation II

$$\begin{aligned}C_n^2 = 1\cdot 10^{-14}(&1899.33 - 28.7688Ws - 96.8552T - 7.63587RH\\&+ 8.03877Ws^2 + 0.600268Ws\,T + 0.0325904Ws\,RH\\&+ 0.0474588T\,RH - 2.16593Ws^3 + 0.0773251T^3\\&+ 0.0028023T\,RH^2 - 0.000533785RH^3 + 0.176089Ws^4\\&- 0.00113842T^4 + 2.61556\cdot 10^{-6}RH^4)\end{aligned} \tag{3.21}$$

3.5.2.3 Model equation III

$$\begin{aligned}C_n^2 = 1\cdot 10^{-14}(&4963.45 - 12.1635Ws - 265.847T - 32.7117RH\\&+ 1.39064Ws^2 + 0.200515Ws\,T + 0.0585552Ws\,RH\\&+ 0.632929T\,RH - 0.167931Ws^3 + 0.213973T^3 + 0.0229636T\,RH^2\\&- 0.00315137RH^3 + 0.00869613Ws^4 - 0.000268105Ws^2\,T^2\\&- 1.55805\cdot 10^{-5}Ws^2\,RH^2 - 0.00311029T^4 - 0.000429867T^2\,RH^2\\&+ 1.38747\cdot 10^{-5}RH)\end{aligned} \tag{3.22}$$

3.5.2.4 Model equation IV

$$\begin{aligned}C_n^2 = 1\cdot 10^{-14}(&3597.95 + 4.81594Ws - 207.508T - 5.633RH\\&- 0.426821Ws^2 - 1.49135Ws\,T + 0.652278Ws\,RH - 0.233119T\,RH\\&+ 0.24708Ws^3 + 0.197403T^3 + 0.0120702T\,RH^2 - 0.00313312RH^3\\&+ 0.0226404Ws^4 - 0.023023Ws^3\,T - 0.000856621Ws^3\,RH\\&+ 0.00521323Ws^2\,T^2 - 0.000181591Ws^2\,RH^2 - 0.00300308T^4\\&+ 5.58919\cdot 10^{-5}T^3\,RH - 0.000192674T^2\,RH^2\\&- 4.38953\cdot 10^{-5}RH^3\,Ws + 4.2236\cdot 10^{-5}RH^3\,T + 1.05632\cdot 10^{-5}RH^4)\end{aligned} \tag{3.23}$$

3.5.2.5 Model equation V

$$\begin{aligned}C_n^2 = 1\cdot 10^{-14}(&5360.63 + 21.0442Ws - 281.763T - 63.5576RH\\&- 0.0431099Ws^2 - 0.101587Ws\,T - 0.271695Ws\,RH\\&+ 2.19559T\,RH - 0.26449Ws^3 + 0.199294T^3 + 0.0168798T\,RH^2\\&+ 0.000579369RH^3 - 0.001449Ws^4 + 0.0101365Ws^3\,T\\&+ 0.00092494Ws^3\,RH - 0.00159949Ws^2\,T^2\\&+ 0.000118693Ws^2\,RH^2 - 0.00265882T^4 - 0.000436822T^3\,RH\\&- 0.000335601T^2\,RH^2 + 7.60425\cdot 10^{-6}RH^3\,Ws\\&- 6.82247\cdot 10^{-5}RH^3\,T + 1.65979\cdot 10^{-6}RH^4)\end{aligned} \tag{3.24}$$

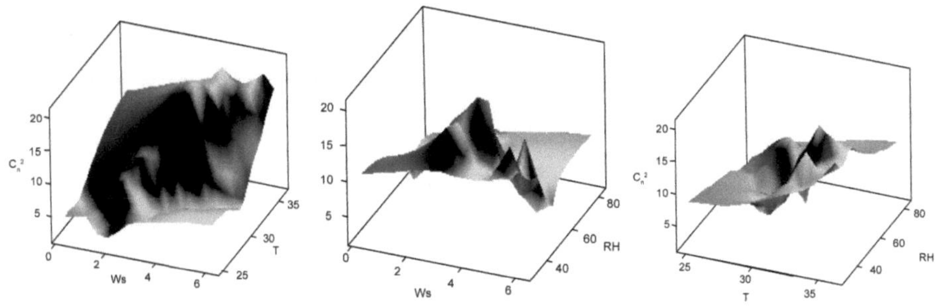

Fig. 3.6: Illustrations of input-output response surface plots. (a) turbulence strength (C_n^2) vs wind speed and temperature, (b) C_n^2 vs wind speed and relative humidity, and (c) C_n^2 vs temperature and relative humidity. Note that the multiplication factor in the z-axis is 10^{-14}.

The individual and combined effects of the input parameters on the performance measures are analyzed and subsequently mathematical models are formulated as in Eqs. (3.20)–(3.24). The sensitivity of the atmospheric parameters on turbulence strength is analyzed. The ANOVA method is used to analyze the individual and combined effects of atmospheric parameters on turbulence strength. Figure 3.6 shows different possible combined effects of wind speed, temperature and relative humidity on turbulence strength.

The computed ANOVA results are used to understand the most dominating input factors on the response surface and only those combinations are considered when optimizing the model. Other regressors are insignificant and their contributions to the square of the model coefficient of determination are negligible. Utmost prediction accuracy is considered while ignoring the less contributing combinations. Many models (Eqs. (3.20)–(3.24)) were examined in order to find a model more suitable for predicting turbulence strength as a function of meteorological parameters. The computed coefficients of determination (R^2) values of proposed turbulence strength prediction model are given against the formulated models in Tab. 3.2. Since the R^2 could Not be Computed (NC) from linear to linear + square + interaction + cubic models for C_n^2, model equation I to model equation V are formulated. As in Tab. 3.2, the model equation V, Eq. (3.24), exhibits the greatest correlation with the measurement values and hence this model alone is considered for practical implementation to predict turbulence strength according to local meteorological data. The residual plots obtained with the considered model, Eq. (3.24), for turbulence strength are shown in Fig. 3.7.

Confirmatory test experiments are conducted using a new set of input conditions to validate the accuracy of the proposed models and a portion of the outcomes of the confirmatory tests are given in Tab. 3.3. The percentage deviation (error) predicted by the proposed models is calculated by

$$\frac{\text{Experimental value} - \text{Predicted value}}{\text{Predicted value}} \cdot 100 \,.$$

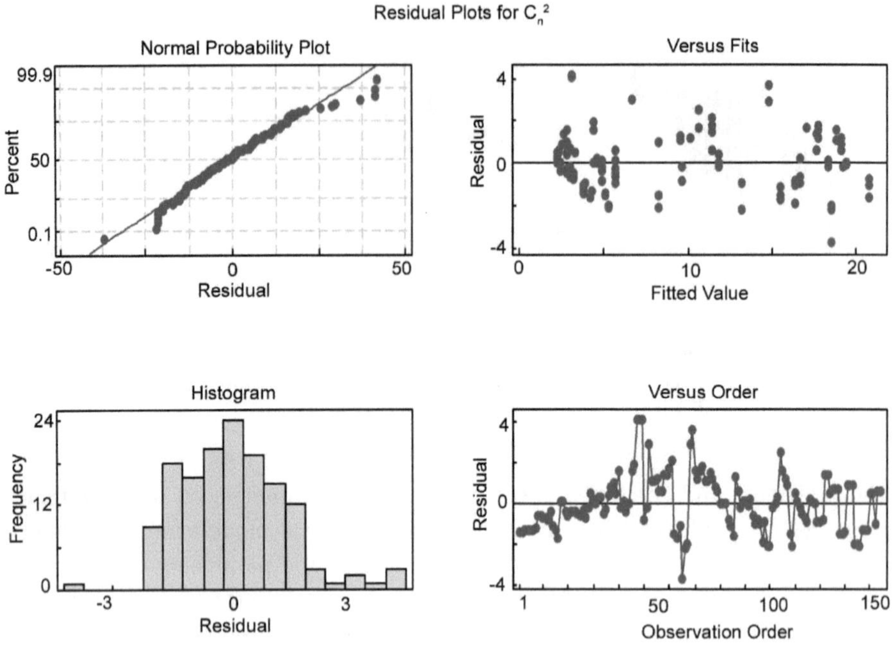

Fig. 3.7: Residual plots of developed regression model (model equation V) for turbulence strength.

Tab. 3.3: Percentage deviations between experimental and prediction values of the proposed models for attenuation and turbulence strength respectively.

Trial ID	Experiment		Prediction		Percentage deviation	
	Attenuation (dB/km)	**C_n^2 $(m^{-2/3}) \cdot 10^{-12}$**	**Attenuation (dB/km)**	**C_n^2 $(m^{-2/3}) \cdot 10^{-12}$**	**Attenuation (%)**	**C_n^2 (%)**
1	3.0931	0.0399	3.0753	0.0393	0.5788	1.5374
2	3.1113	0.0481	3.0753	0.0441	1.1706	9.0497
3	3.0810	0.0483	3.0753	0.0441	0.1853	9.4473
4	3.1275	0.0608	3.0753	0.0579	1.6974	5.0417
5	3.0978	0.0816	3.0642	0.0773	1.0965	5.5858
6	3.0818	0.0774	3.0642	0.0773	0.5744	0.2003
7	3.0851	0.1044	3.0642	0.1028	0.6821	1.6269
8	3.8964	0.1038	3.8059	0.1028	2.3779	1.0333
9	3.8734	0.1065	3.8059	0.1028	1.7736	3.6234
10	3.8527	0.0399	3.8059	0.0393	1.2297	1.5374
			Average percentage deviation		**1.13663**	**3.86832**

The individual and average percentage deviation results confirm the suitability of the proposed models at our test field.

3.6 Experimental results and data analysis

The measured meteorological and visibility data are applied as inputs to the models discussed in Sections 3.4 and 3.5 and the prediction accuracy is investigated intensively. Some of the important data experimentally acquired from the real-world FSO link in different local seasons: monsoon, rainy, winter, presummer and summer and in a one year period are presented and from them the potential and feasibility of all the selected and proposed models are evaluated. The diurnal period performance of meteorological parameters is studied in terms of minimum and maximum (min, max) values and standard deviation (SD). The outstanding performance of the proposed models in predicting atmospheric attenuation and turbulence strength are analyzed and the results are presented in this section.

3.6.1 Comparison of the predicted attenuation data with measured values

The local meteorological and visibility (using Eq. (3.1)) data are measured for every second in the diurnal periods, averaged data are recorded and the environmental changes are studied. The meteorological and visibility data are substituted in Eqs. (3.3), (3.4), (3.6), (3.7), and (3.19) as required and prediction results are compared against the measured attenuation using Eq. (3.2). The selected and proposed models are implemented in the MATLAB environment. The accuracies of the models' predictions are carefully analyzed in terms of RMSE. The profile of recorded real-time meteorological and visibility data and prediction and measurement values of optical attenuation are described below for five days chosen somewhat randomly in different local seasons (since FSOC is mainly seasonally dependent) in a one year period.

3.6.1.1 Data for 6th June 2013, monsoon

The weather parameters varied from 1 m s^{-1} to 7.2 m s^{-1} with SD of 1.59 m s^{-1} for wind speed, 27 °C to 39 °C with SD of 3.75 °C for temperature, 25 % to 84 % with SD of 15.89 % for relative humidity and 2 km to 10 km with SD of 2.19 km for visibility as shown in Fig. 3.8 (a–d). SD of RH, T, and V are high while Ws is low. The different atmospheric conditions observed on 6th June 2013 (Thursday) are scattered cloudy, partial cloudy and mostly cloudy, light drizzly, light rainy and hazy.

Figure 3.9 (a) shows the experimental time series plot of attenuation predicted from local meteorological and visibility data collected on 6th June 2013 and measured values. The max-min values of measured attenuation are ≈ 6.45 dB/km and 1.74 dB/km at about 4.00 pm and 10.00 pm respectively with SD of 1.29 dB/km. A short interval of high attenuation ≈ 6.45 dB/km is seen in the time span 3.00 pm to 4.00 pm due to the normal wind speed and high temperature and low relative humidity (Font et al. 2006). In Fig. 3.9, the colors cyan, black, red, blue, green and magenta are used to represent

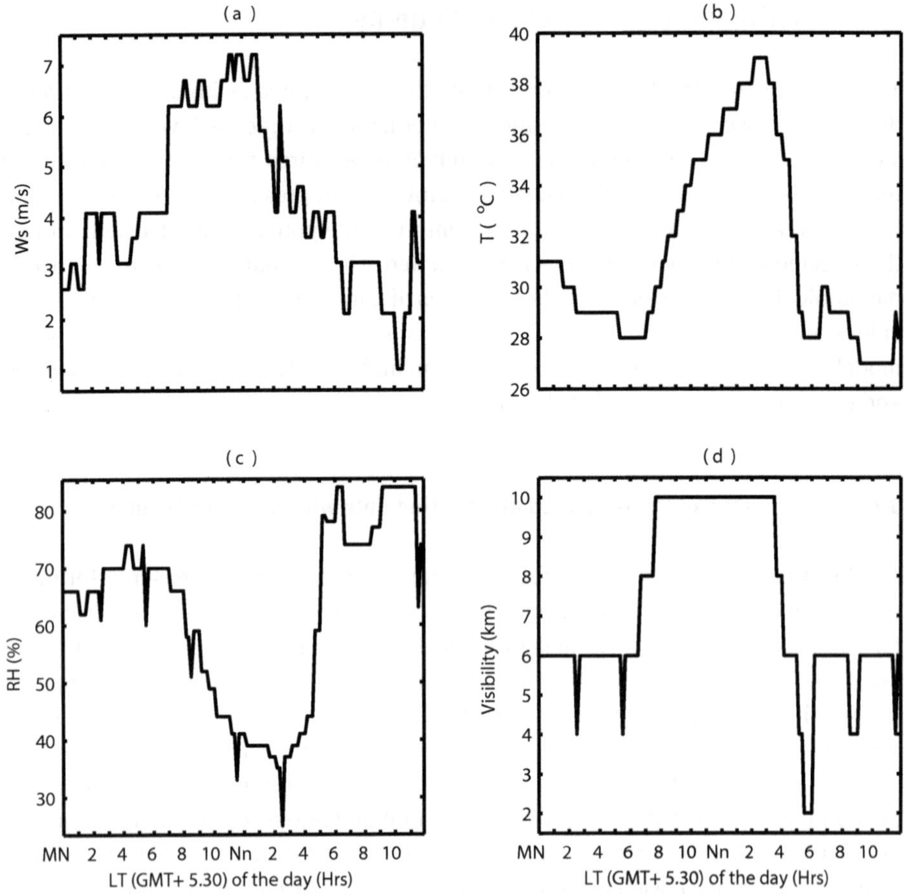

Fig. 3.8: Diurnal time series profile of measured macrometeorological parameters (a–c) and visibility (d). MN: midnight and Nn: noon.

the measurement, Ijaz, Awan, Itai Drod, Bataille and proposed models respectively. Comparing the selected models, the prediction values of attenuation obtained using M. Ijaz's and M. S. Awan's models greatly differ from the measured values with the maximum deviations of 4.54 dB/km and 6.24 dB/km respectively as in Fig. 3.9 (a) and are not considered to be accurate at our test field. This is because these models were developed for specific situations with small data sets, therefore, they often failed when applied to the real-world open atmosphere (Dror et al. 1996). M. Ijaz's model gives low and high attenuation of ≈ 1.68 dB/km and ≈ 8.40 dB/km with SD of ≈ 1.19 dB/km when the visibility is high: 10 km and low: 2 km in the time spans 8.00 am to 3.00 pm and 5.05 pm to 6.03 pm respectively. Reasonable correlation is seen at an average value of 2.81 dB/km at about MN to 6.30 am in this season. However, the pattern of exhibiting less correlation and disagreement with the measurement data over longer periods and data sets are observed. These results suggest that the sole dependency on visibility as in Eq. (3.3) is insufficient and the Ws, T, and RH are also seen to exhibit statistical significance and they are interwoven with the dependence in atmospheric attenuation effects (Ijaz et al. 2013b).

The min/max values of attenuation predicted by M. S. Awan's model are $5.58 \cdot 10^{-6}$ dB/km and 39.62 dB/km with SD of 12.51 dB/km at about 2.30 pm and around 10.00 pm respectively when the RH is very low: 25 % and high: 84 %. This statistics are inappropriate and the model exhibits adequate deviations from another models and measured values as well. Furthermore, Fig. 3.9 (a) indicates that the relationship of RH to attenuation as in Eq. (3.4) is not correct for the measurements at our test field, since the variation's band of attenuation is large and does not correlate with meteorological fluctuations. Although the relative humidity is the only variable in this model, the results as in Fig. 3.9 (a) shows here that the Ws and T should also be considered to obtain even a reasonable prediction since they also have an effect in the open atmosphere. Therefore, this model is unfit to our test field and it is clear that there is no one-to-one relationship as already reported in Bouchet et al. (2006), Dror et al. (1996), Randall (2004) and Font et al. (2006). It is clear from Eq. (3.4) and Fig. 3.9 (a) that relative humidity alone cannot be used reliably to predict atmospheric attenuation.

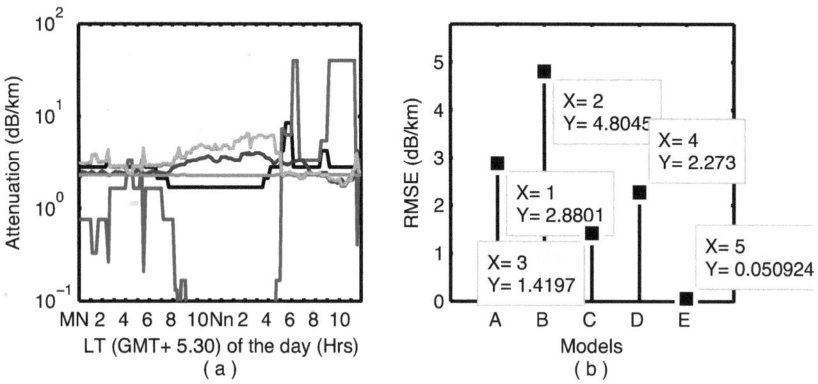

Fig. 3.9: Graphical comparison of time series profile of atmospheric attenuation predicted using selected existing models and proposed models and measured from direct beam transmission signal power (a). Root mean square error value against the particular model (b).

The min/max values of attenuation predicted by the Itai Dror's model, as in Eq. (3.6), are 1.88 dB/km and 3.97 dB/km with SD of 1.44 dB/km at about 8.00 am to 4.30 pm and 3.00 pm to 10.15 pm respectively when the Ws, T are high: 7.2 m s^{-1}, 38 °C; RH is low: 37 % and Ws, T are low: 2.1 m s^{-1}, 27 °C; RH is high: 84 %. These statistics significantly differ from the measured values. In our experiments, the prediction results show good agreement and correlation with the measured data during MN to morning and evening to MN since the Ws is low: 2.6 m s^{-1}, T is low: 28 °C and RH is relatively high: 76 %. However, the results obtained in the daytime disagreed with less contradiction to the measured values due to the model inputs limitations and unavailability of regression coefficients for 850 nm. Therefore, this model could not yield an accurate estimation of atmospheric attenuation at our test field. The min/max values of the Bataille are 2.25 dB/km and 2.29 dB/km with SD of 0.007 dB/km. These results repeat themselves and exhibit almost singularity at every time of day with an average constant attenuation of 2.27 dB/km. This behavior clearly shows that this model as in Eq. (3.7) is unsuitable to our test field in this season. Although generally there is no agreement with the measurements, there are certain instance when the prediction results reasonably match the measurements over short periods of time, for example, as in Fig. 3.9 (a), from 5.00 pm to 9.00 pm when the Ws, T are low: 2.1 m s^{-1}, 28 °C and RH is high: 75 %.

The min/max values of the proposed model are 1.74 dB/km and 6.40 dB/km with SD of 1.29 dB/km at about 9.00 pm to 11.15 pm and 9.00 am to 5.00 pm respectively when the Ws and T are low: 1 m s^{-1}, 28 °C; RH is high: 84 % and Ws, T are high: 7.2 m s^{-1}, 38 °C; RH is low: 35 %. The values predicted using the proposed model, as in Eq. (3.19), could yield a good correlation to the measured values at our test field and the results reveals that the proposed model exhibits best prediction with the deviation error of ≈ 0 dB/km throughout the diurnal periods of experimentation. The prediction accuracy is validated by computing the RMSE using the predicted and measured values. The results show as in Fig. 3.9 (b) that the proposed model gives a much lower RMSE value of 0.04 dB/km whereas other models are 2.87 dB/km, 4.80 dB/km, 1.41 dB/km, and 2.27 dB/km.

3.6.1.2 Data for 2nd November 2013, rainy

The weather parameters varied from 0.27 m s^{-1} to 4.10 m s^{-1} with SD of 1.29 m s^{-1} for wind speed, 25 °C to 32 °C with SD of 1.95 °C for temperature, 54 % to 94 % with SD of 10.58 % for RH and 1 km to 5 km with SD of 1.05 km for visibility as shown in Fig. 3.10 (a–d). The SD of RH is high while Ws, T, and V are low. The different atmospheric conditions observed on 2nd November 2013 (Saturday) are hazy, rainy, light rainy, thunderstorms and rainy, light thunderstorms and rainy, light thunderstorms and smoky.

Figure 3.11 (a) shows the experimental time series plot of attenuation predicted from local meteorological and visibility data collected on 2nd November 2013 and measured values. The min/max values of measured attenuation are ≈ 1.64 dB/km and ≈ 4.91 dB/km at about 2.00 pm to 8.30 pm and Nn respectively with SD of 0.69 dB/km. A short interval of high attenuation ≈ 4.91 dB/km is seen in the time span of 9.30 am to Nn due to the high Ws, T: ≈ 4.1 m s^{-1}, 32 °C and very low RH: ≈ 54 % on 2nd November, 2013. Comparing the selected models: The min/max values of attenuation predicted by M. Ijaz's model are ≈ 3.36 dB/km and ≈ 16.82 dB/km with SD of ≈ 2.24 dB/km when V is high: ≈ 5 km and low: ≈ 1 km in the time span of MN to 5.30 am and 8.30 pm to 10.00 pm

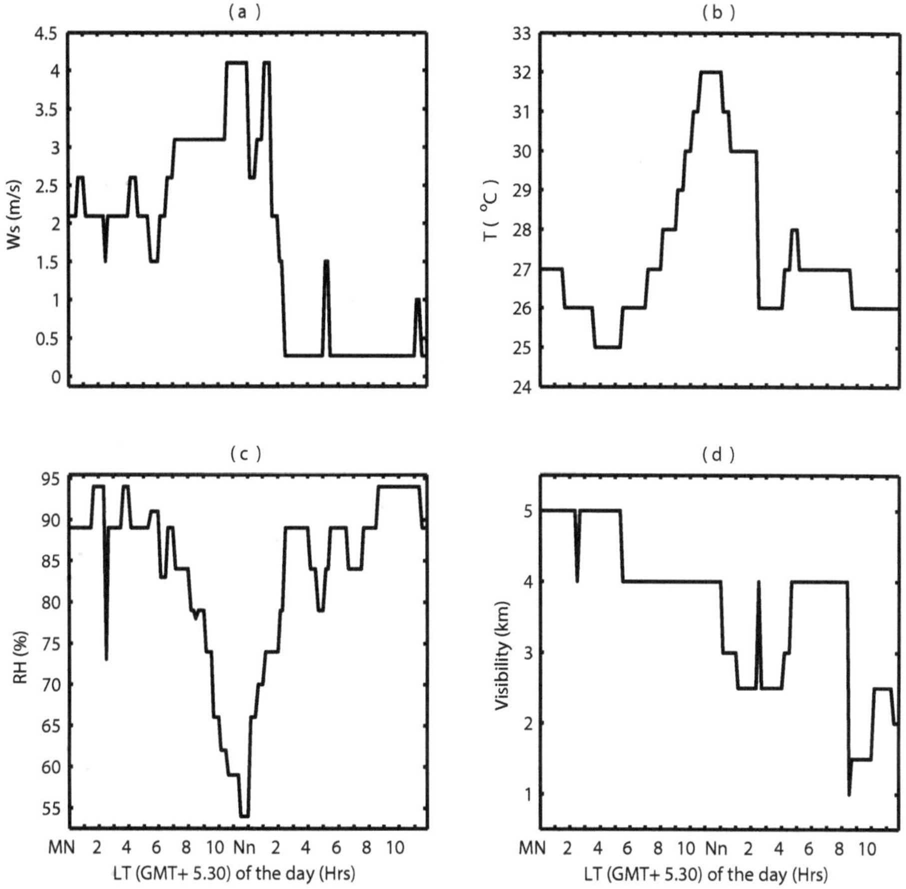

Fig. 3.10: Diurnal time series profile of measured macrometeorological parameters (a–c) and visibility (d). MN: midnight and Nn: noon.

respectively. This model prediction result always falls above the measurements with a maximum deviation of 14.97 dB/km. The prediction result's deviation is relatively small: ≈ 1.7 dB/km from MN to 5.00 am and then the deviation significantly increases i.e. the result recedes from the measurements due to the V variations: ≈ 5 km to 1 km as in Fig. 3.10 (d) and 3.11 (a). However, the prediction results meet the measurements in four instances: 2.30 am, 4.30 am, 11.15 am, and around 11.45 am at the prediction values of 3.31 dB/km, 3.30 dB/km, 3.98 dB/km, and 4.8 dB/km respectively. Most of the time the prediction values are away from the measured values.

The min/max values of attenuation predicted by M. S. Awan's model are 0.04 dB/km and ≈ 41.06 dB/km with SD of ≈ 14.66 dB/km at about 11.45 am to MN and around 2.00 am, 4.00 am, 8.30 am, and 3.45 pm to MN respectively as in Fig. 3.11 (a), when the RH is very low: 54 % and very high: 94 % as in Fig. 3.10 (d). The prediction results always fall above the measurements and fall below for certain courses of time from 11.00 am to 3.00 pm. The prediction results linearly depend on the RH and mostly fluctuate within the smaller range of ≈ 25.65 dB/km and ≈ 41.06 dB/km before 8.00 am and after 3.00 pm as in Figs. 3.10 (c) and 3.11 (a). A rapidly increasing and decreasing attenuation trend is seen, like the RH measured on 2nd November 2013 from 8.30 am to 2.30 pm. However, the predication results greatly differ from the measurements with an average attenuation of 23.51 dB/km.

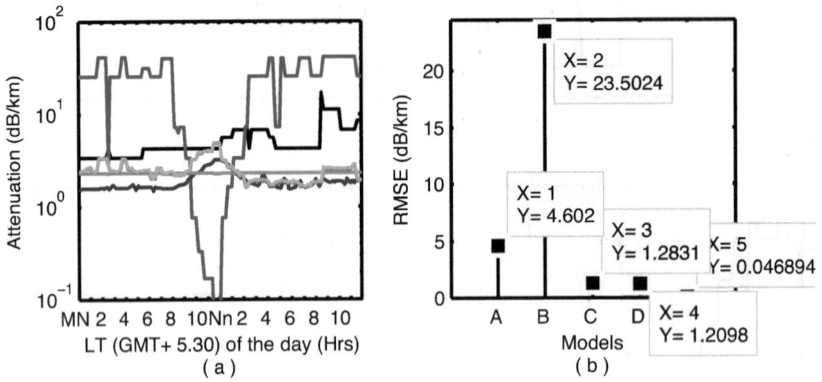

Fig. 3.11: Graphical comparison of time series profile of atmospheric attenuation predicted using selected existing models and proposed models and measured from direct beam transmission signal power (a). Root mean square error value against the particular model (b).

The min/max values of attenuation predicted by Itai Dror's model are: 1.47 dB/km and 3.16 dB/km with SD of 0.39 dB/km about MN to 10.00 am and 10.00 am to 12.15 pm respectively when the Ws, T are low: 2.1 m s^{-1}, 27 °C; RH is high: ≈ 90 % and Ws, T are high: 4.1 m s^{-1}, 32 °C; RH is low: 54 %. The prediction results go to infinite from MN to 8.00 am and 3.00 pm to MN due to the limitations of the model i.e. RH appears above 85 % which falls out of the range limitations (Dror et al. 1996). Therefore, the previous i.e. $(n-1)^{th}$ sample is taken into account instead of infinite for computation convenience as in Fig. 3.11 (a) and the min/max values are computed. The prediction accuracy increases and becomes close to measurements as the inputs fall with the limitation ranges, for example, during 8.00 am to 3.00 pm in Fig. 3.11 (a). Note that most of the time metrological parameters fall out of the limitation range at our test field. The Bataille model's min/max values are 2.28 dB/km and 2.30 dB/km with SD of 0.004 dB/km. The model predicts an average constant value of 2.29 dB/km at almost all times of day. The prediction values reasonably correlate with measurements during MN to 9.30 am and 8.30 pm to 11.30 pm with maximum deviation of 2.52 dB/km. When the RH is high: 89 %, T is low: 27 °C. This trend is deviated from when the RH is low: 54 % and T is high: 32 °C. The min/max values of the proposed model are

1.60 dB/km and 4.82 dB/km with SD of 0.60 dB/km at about 2.30 pm to 8.30 pm and around 11.30 pm and 11.00 am to 12.00 am respectively when the Ws, T are very low: 0.27 m s^{-1}, 26 °C; RH is very high: 90 % and Ws, T are high: 4.1 m s^{-1}, 32 °C; RH is low: 54 %. The values predicted using the proposed model could yield a good correlation to the measurement values. The computed RMSE results in Fig. 3.11 (b) show that the proposed model gives a far lower RMSE value of 0.04 dB/km whereas other models give 4.60 dB/km, 23.50 dB/km, 1.28 dB/km, and 1.20 dB/km.

3.6.1.3 Data for 15th December 2013, winter

The weather parameters varied from 1.5 m s^{-1} to 5.1 m s^{-1} with SD of 0.97 m s^{-1} for wind speed, 23 °C to 31 °C with SD of 2.65 °C for temperature, 52 % to 94 % with SD of 13 % for relative humidity and 1 km to 4 km with SD of 0.87 km for visibility as shown in Fig. 3.12 (a–d). The SD of the T and RH is high while Ws and V are low. The different atmospheric conditions observed on 15th December 2013 (Sunday) are hazy, scattered clouds, mist, fog and partially cloudy.

Figure 3.13 (a) shows the experimental time series plot of attenuation predicted from local meteorological and visibility data collected on 15th December 2013 and measured values. The min/max values of measured attenuation are ≈ 1.99 dB/km and 5.60 dB/km at about MN and 1.30 pm to 2.15 pm respectively with SD of 0.80 dB/km. A short interval of high attenuation ≈ 5.60 dB/km is seen around 2.00 pm due to the high Ws, T: 4.1 m s^{-1}, 31 °C and very low RH: 55 %. Comparing the selected models: the M. Ijaz model's low and high attenuations are ≈ 4.20 dB/km and ≈ 16.82 dB/km with SD of ≈ 3.58 dB/km when V is high: 4 km and low: 1 km in the time span of 2.15 pm to 6.30 pm and 5.00 am to 6.30 am respectively. This model prediction result exhibits an approximately increasing and decreasing trend with the maximum deviation of 14.62 dB/km from MN to 8.30 am as in Fig. 3.13 (a), since V exhibits a decreasing and increasing trend as in Fig. 3.12 (d). The prediction and measurement meet each other at a value of 3.97 dB/km around 3.00 pm when Ws is normal: 3.1 m s^{-1}, T is high: 30 °C and RH is low: 62 %. Apart from this instant, the prediction values always fall above the measurements. The min/max values of attenuation predicted by M. S. Awan's model are 0.02 dB/km and 41.06 dB/km with SD of 15.17 dB/km at about 2.45 pm and 3.45 am to 8.00 am respectively when the RH is very low: 52 % and very high: 94 %. This model prediction result fluctuates largely with a maximum deviation of 37.76 dB/km. The prediction results do not correlate with the measurements except two in instances: 3.15 dB/km and 2.88 dB/km at around 9.00 am and 7.15 pm respectively, as shown in Fig. 3.13 (b), since RH exhibits a decreasing and increasing trend as in Fig. 3.12 (c).

The min/max values of attenuation predicted by Itai Dror's model are 1.25 dB/km and 2.92 dB/km with SD of 0.48 dB/km about MN to 8.30 am and 1.15 pm respectively when the Ws, T are low: 2.6 m s^{-1}, 24 °C; RH is high: 94 % and Ws, T are high: 5 m s^{-1}, 30 °C; RH is low: 57 %. The model yields the constant response from MN to 8.00 am, since the RH falls above 85 % as in Fig. 3.12 (c), i.e. beyond the input limitation. The

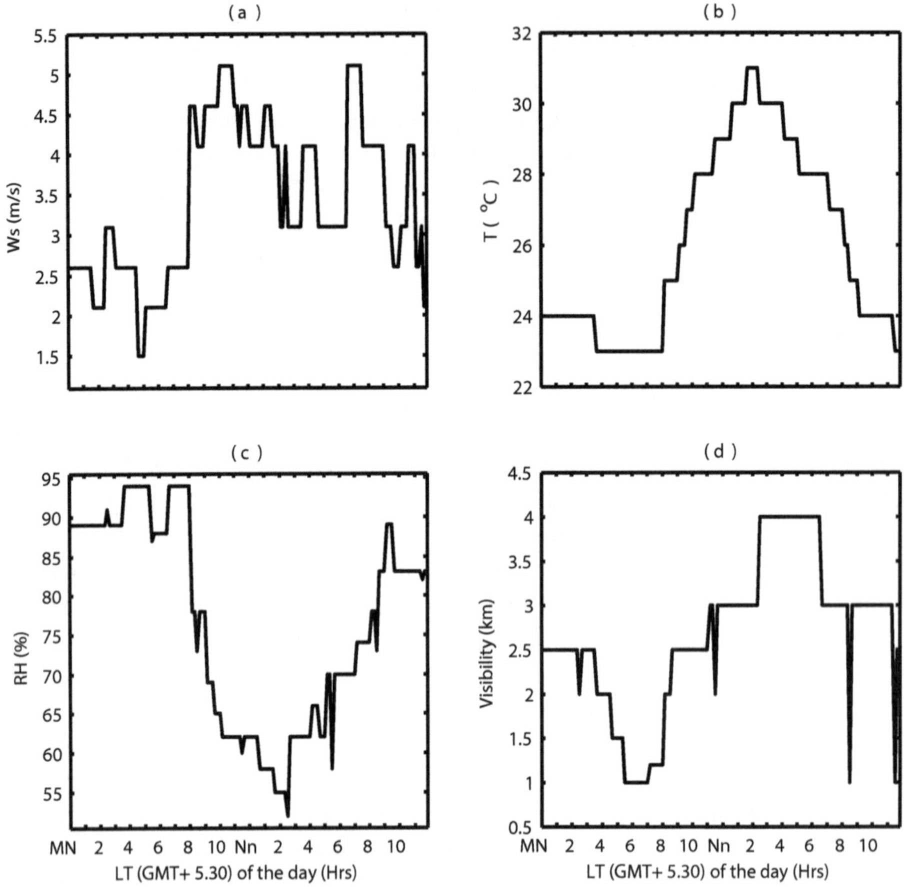

Fig. 3.12: Diurnal time series profile of measured macrometeorological parameters (a–c) and visibility (d). MN: midnight and Nn: noon.

model gives reasonable results once RH falls under 85 % as in Fig. 3.13 (a) and increases to maximum, since RH comes down to 52 %, T goes up to 31 °C and a highly fluctuating trend is seen in Ws. The maximum deviation of prediction is 2.92 dB/km and approximately moves along with measurements from 7.30 pm to MN, since Ws is normal: ≈ 3.1 m s^{-1}, T is low: ≈ 24 °C and RH is high: ≈ 83 %. The Bataille model's min/max values are 2.27 dB/km and 2.28 dB/km with SD of 0.0035 dB/km. The model predicts an average constant value of 2.28 dB/km. Both Itai Dror's and this model predict almost similar results from 10.30 am onwards. The maximum deviation is 2.87 dB/km i.e. these two models exhibit an almost equal magnitude of deviation from the measurement values. The min/max values of the proposed model are 1.94 dB/km and 5.59 dB/km with SD of 0.81 dB/km about MN and Nn to 4.05 pm respectively when the Ws, T are high: 4.6 m s^{-1} and 31 °C and RH is very low: 52 %. The RMSE results in

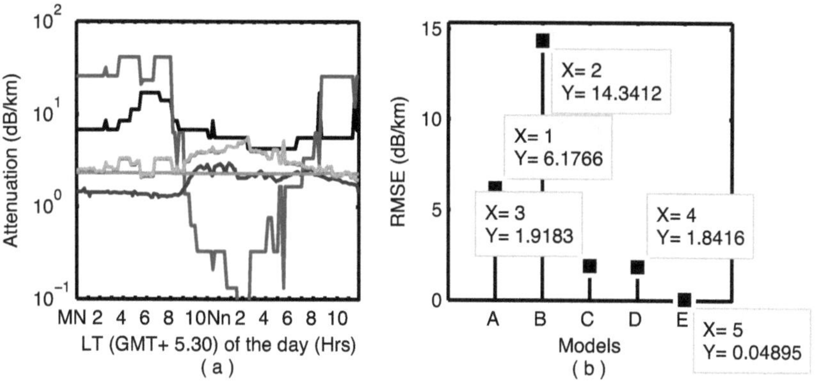

Fig. 3.13: Graphical comparison of time series profile of atmospheric attenuation predicted using selected existing models and proposed models and measured from direct beam transmission signal power (a). Root mean square error value against the particular model (b).

Fig. 3.13 (b) show that the proposed model gives a far lower RMSE value of 0.05 dB/km whereas the other models give 6.17 dB/km, 14.33 dB/km, 1.92 dB/km, and 1.84 dB/km.

3.6.1.4 Data for 14th March 2014, presummer

The weather parameters varied from 0.27 m s^{-1} to 7.2 m s^{-1} with SD of 2.08 m s^{-1} for wind speed, 21 °C to 35 °C with SD of 4.62 °C for temperature, 16 % to 88 % with SD of 20.75 % for relative humidity and 2 km to 8 km with SD of 1.87 km for visibility as shown in Fig. 3.14 (a–d). The SD of Ws, T, and RH are high while V is low. The different atmospheric conditions observed on 14th March 2014 (Friday) are partially cloudy and hazy.

Figure 3.15 (a) shows the experimental time series plot of attenuation predicted from local meteorological and visibility data collected on 14th March 2014 and measured values. The min/max values of measured attenuation are ≈ 1.63 dB/km and 8.82 dB/km at about 2.15 am to 8.00 am and around Nn respectively with SD of 1.94 dB/km. A short interval of high attenuation ≈ 8.82 dB/km is seen due to the high Ws, T: 5.1 m s^{-1}, 30 °C and very low RH: 48 %. The M. Ijaz model's low and high attenuations are ≈ 2.10 dB/km and ≈ 8.40 dB/km with SD of ≈ 2.09 dB/km when V is high and low: 8 km and 2 km respectively in the time span of 2.30 pm to 6.30 pm and 5.30 am to 9.00 am respectively as in Figs. 3.14 (d) and 3.15 (a). The prediction result appears above the measurements up to 10.30 am and then falls below with the maximum deviation of 6.68 dB/km, since V falls down to 2 km, then increases to a maximum of 7.2 km. The correlation between prediction and measurements is very poor except for one instant at 5.06 dB/km at around 10.30 am, when V reaches 4 km with a increasing trend. The min/max values of attenuation predicted by M. S. Awan's model are $1.61 \cdot 10^{-7}$ dB/km and 25.65 dB/km with SD of 10.08 dB/km at about 5.15 pm and 2.15 am to

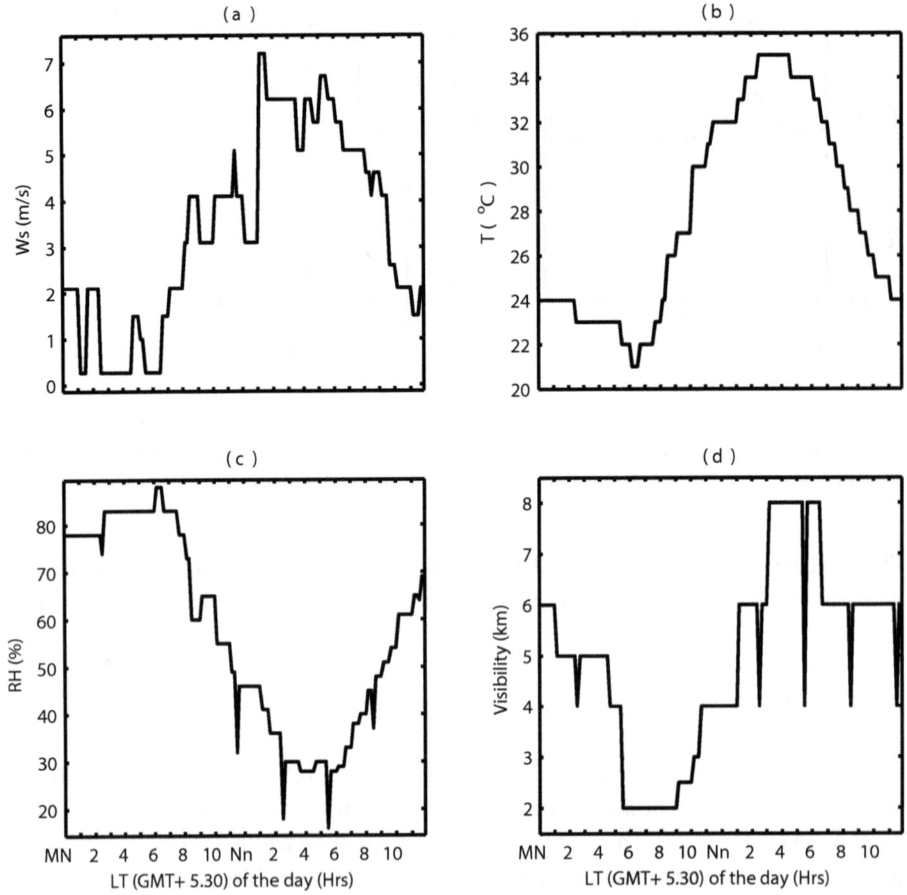

Fig. 3.14: Diurnal time series profile of measured macrometeorological parameters (a–c) and visibility (d). MN: midnight and Nn: noon.

7.10 am respectively when RH is very low: 16 % and high: 88 %. This model's results almost fall below the measurements with the maximum deviation of 25.64 dB/km as in Fig. 3.15 (a). The prediction values are above the measurements as long as RH is high: 83 %. The predicted attenuation exhibits a decreasing and increasing trend as the variations in RH.

The min/max values of attenuation predicted by Itai Dror's model are 1.97 dB/km and 5.50 dB/km with SD of 1.02 dB/km at about 11.00 am to 6.20 pm and MN respectively when the Ws, T are low: 2 m s^{-1}, 24 °C; RH is high: 83 % and Ws, T are high: 6.3 m s^{-1}, 35 °C; RH value is very low: 36 %. The model response follows the measurements with an approximate constant deviation of 3.2 dB/km, since Ws and T keep roughly a smooth increasing and decreasing trend while RH keeps a decreasing and increasing trend as in Figs. 3.14 (a–c) and 3.15 (a). The predicted and measured values

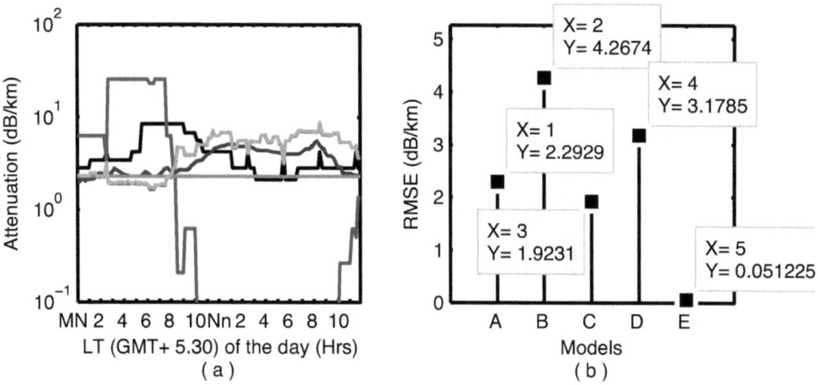

Fig. 3.15: Graphical comparison of time series profile of atmospheric attenuation predicted using selected existing models and proposed models and measured from direct beam transmission signal power (a). Root mean square error value against the particular model (b).

are equal at 2.76 dB/km, 4.76 dB/km, and 5.58 dB/km at around 08.00 am, 1.05 pm, and 3.10 pm. The Bataille model's min/max values are 2.23 dB/km and 2.27 dB/km with SD of 0.0088 dB/km. The average constant value of 2.269 dB/km is maintained throughout the period of measurement. In this season, the model prediction widely deviated from the measurements from 8.30 am onwards and was merely close to M. Ijaz's model from 1.00 pm onwards. This model holds a good correlation with the measurements at 2.33 dB/km at around 2.00 am and then increases with maximum deviation of 4.74 dB/km. The min/max values of the proposed model are 1.63 dB/km and 8.76 dB/km with SD of 1.94 dB/km at about MN and 9.00 am to 11.45 pm respectively when the Ws, T are very low: 2.1 m s^{-1}, 24 °C; RH is high: 83 % and vice versa respectively. The results shown in Fig. 3.15 (b) prove that the proposed model gives a far lower RMSE value of 0.054 dB/km whereas the other models give 2.29 dB/km, 4.264 dB/km, 1.92 dB/km, and 3.18 dB/km.

3.6.1.5 Data for 12th May 2014, summer

The weather parameters varied from 1 m s^{-1} to 7.4 m s^{-1} with SD of 1.43 m s^{-1} for wind speed, 28 °C to 40 °C with SD of 4.36 °C for temperature, 26 % to 84 % with SD of 16.7 % for relative humidity and 2 km to 8 km with SD of 1.57 km for visibility as shown in Fig. 3.16 (a–d). The SD of Ws, T, RH, and V are high. The different atmospheric conditions observed on 12th May 2014 (Monday) are hazy, partial cloudy and light thunderstorms.

Figure 3.17 (a) shows the experimental time series plot of attenuation predicted from local meteorological and visibility data collected on 12th May 2014 and measured values. The min/max values of measured attenuation are ≈ 1.64 dB/km and 8.11 dB/km at about MN to 9.00 am and Nn to 4.30 pm respectively with SD of 2.05 dB/km. A short

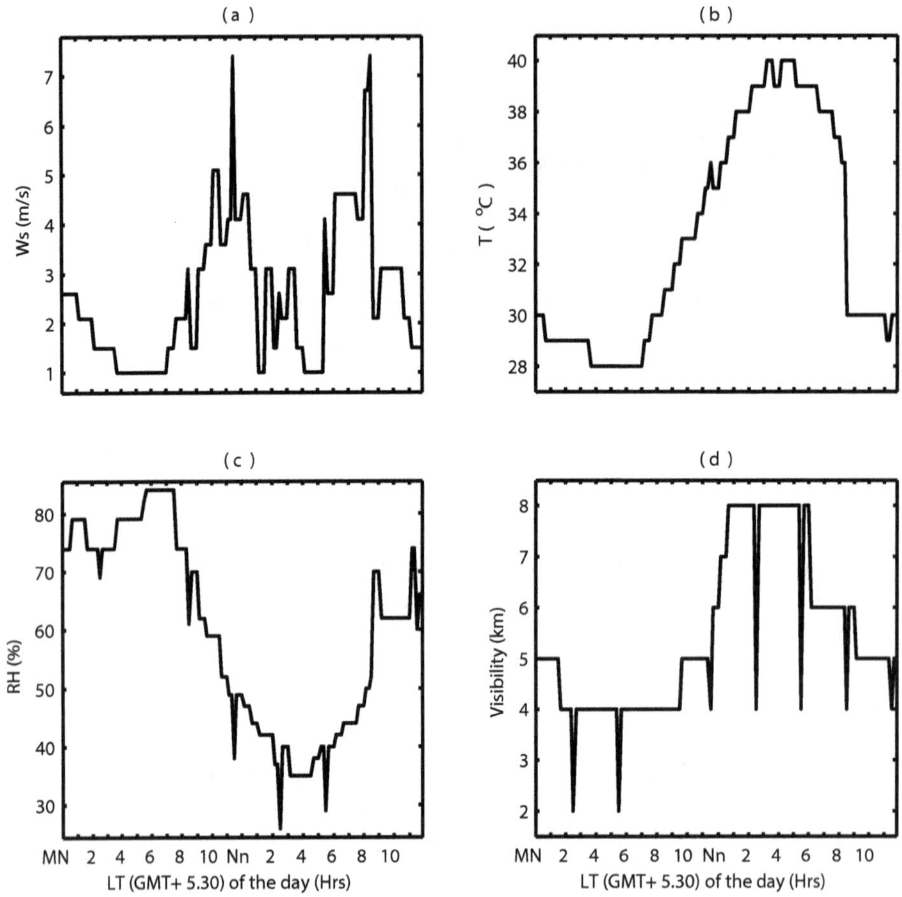

Fig. 3.16: Diurnal time series profile of measured macrometeorological parameters (a–c) and visibility (d). MN: midnight and Nn: noon.

interval of high attenuation ≈ 7.9 dB/km is seen at low Ws: 1.8 m s^{-1}; high T: 39 °C and very low RH: 38 %.

Comparing the selected models: the M. Ijaz model's low and high attenuation are ≈ 2.10 dB/km and ≈ 8.40 dB/km with SD of ≈ 1 dB/km when visibility is high: 8 km and low: 2 km in the time span from 1.00 pm to 5.30 pm and 2.30 am to 5.30 am respectively. About 9.30 am, the prediction results fall above and below the measurements with the maximum deviation of 5.42 dB/km. Figures 3.15 (a) and 3.17 (a) clearly show that this model exhibits a resemblance to presummer in this season as well.

The min/max values of attenuation predicted by M. S. Awan's model are $8.10 \cdot 10^{-6}$ dB/km and 39.62 dB/km with SD of 10.81 dB/km at about 5.30 am to 6.30 am and around 2.30 pm respectively when the RH is very low: 26 % and high: 84 %. This model prediction values vary greatly with a deviation of 37.95 dB/km. The model exhibits very minimum values for the maximum duration of experimentation especially in the daytime from 8.00 am to 6.30 pm as in Fig. 3.17 (a) and keeps very poor correlation. The prediction results meet the measurements at 2.15 dB/km at around 8.17 am when T and RH meet each other in the increasing and decreasing trend respectively as in Fig. 3.16 (b) and (c). The min/max values of attenuation predicted by the Itai Dror model are 1.76 dB/km and 5.24 dB/km with SD of 0.91 dB/km at about MN and Nn to 4.30 pm respectively when the Ws, T are low: 2.1 m s^{-1}, 29 °C and RH is high: 80 %; Ws, T are high: 4.6 m s^{-1} and 40 °C; RH is low: 40 % respectively This model also shows similar behavior in this season as to presummer. The prediction results are almost close to the measurements at ≈ 2.21 dB/km at about MN to 7.30 am when the Ws, T are low: ≈ 1.5 m s^{-1}, 28.6 °C and RH is high: 81 % as was the case in the previous season. Apart from 7.30 am, the prediction differs with a maximum deviation of 3.14 dB/km as in Fig. 3.17 (a). Starting from 10.00 am, the prediction once again becomes approximately equal to measurements at 4.13 dB/km since the Ws, T are low: 2.1 m s^{-1}, 29 °C; RH is high: 74 %.

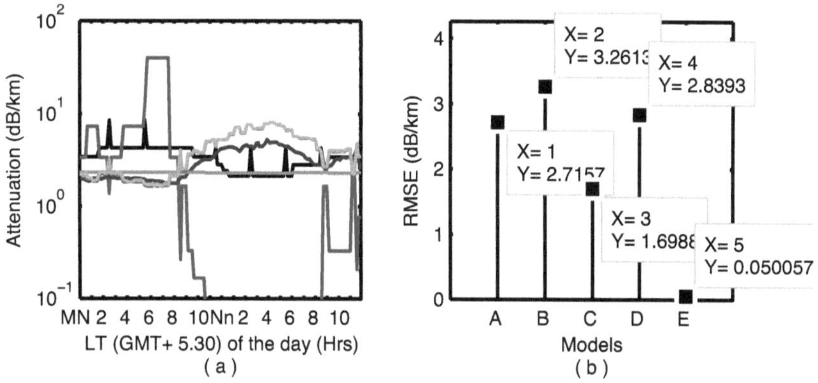

Fig. 3.17: Graphical comparison of time series profile of atmospheric attenuation predicted using selected existing models and proposed models and measured from direct beam transmission signal power (a). Root mean square error value against the particular model (b).

The Bataille model's min/max values are 2.261 dB/km and 2.302 dB/km with SD of 0.0069 dB/km. The average constant value of 2.288 dB/km is maintained throughout the period of measurement. This model works well in the nighttime in this season when the T is low: 29 °C and RH is high: 79 % at about MN to 3.00 am. The Itai Dror and this model's prediction values keep close correlation with the measurements from MN to 9.00 am due to the calm atmospheric conditions. The deviation in the day time is greater with a maximum of 5.34 dB/km as in Figs. 3.16 (a–c) and 3.17 (a) and the deviation decreases from 8.30 pm onwards. The min/max values of the proposed model are 1.64 dB/km and 8.03 dB/km with SD of 2.05 dB/km at about MN to 9.30 am and

11.30 am to 4.00 pm respectively when the Ws, T are very low: 2 m s^{-1}, 29 °C; RH is high: 80 % and Ws is highly sporadic: 1.5 m s^{-1} to 7.4 m s^{-1}; T is high: 40 °C; RH is very low: 32 %. The results shown in Fig. 3.17 (b) exhibit that the proposed model gives a far lower RMSE value of 0.0503 dB/km whereas other models give 2.71 dB/km, 3.26 dB/km, 1.69 dB/km, and 2.83 dB/km.

3.6.2 Comparison of predicted C_n^2 data with measured values

The necessary data and measured values of meteorological, geographical, geometrical and received beam parameters are substituted in Eqs. (3.8)–(3.11), (3.13), (3.14), and (3.24) as required. The prediction results of C_n^2 are recorded and compared against the measured (Ni, 2009) C_n^2 values. The accuracy of the models' prediction are carefully analyzed in terms of Sum of Absolute Error (SAE). The analysis on the profile of real-time meteorological data predicted and measured C_n^2 values for five different days in a one year period are presented below.

3.6.2.1 Data for 16th March 2013, presummer

The weather parameters varied from 0.27 m s^{-1} to 6.18 m s^{-1} with SD of 1.50 m s^{-1} for wind speed, 25 °C to 36.11 °C with SD of 3.60 °C for temperature, 30 % to 83 % with SD of 14.64 % for relative humidity and 101.1 kPa to 101.6 kPa with SD of 0.14 kPa for pressure (P) as shown in Fig. 3.18 (a–d). The SD of T and RH are high; Ws and P are relatively low. The different atmospheric conditions observed on 16th March 2013 (Saturday) are partially cloudy, hazy, scattered cloudy and mostly cloudy. Uneven Ws and RH variations are observed, as shown in Fig. 3.18 (a–c).

Figure 3.19 (a) shows the experimental time series plot of C_n^2 predicted from local meteorological data collected on 16th March 2013 and measured values. The min/-max values of measured C_n^2 are $\approx 2.36 \cdot 10^{-14}\,m^{-2/3}$ and $2.01 \cdot 10^{-13}\,m^{-2/3}$ about early morning and Nn respectively with SD of $5.90 \cdot 10^{-14}\,m^{-2/3}$ when Ws, T are low: 0.27 m s^{-1}, 25 °C and RH is high: 83 %. In Fig. 3.19, the colors cyan, black (A), red (B), blue (C), green (D), and magenta (E) are used to represent the measurement, PAMELA, Hufnagel–Valley, beam wandering, polynomial regression, and proposed models respectively.

The min/max values of C_n^2 predicted by the PAMELA model are $5.46 \cdot 10^{-20}\,m^{-2/3}$ and $2.23 \cdot 10^{-10}\,m^{-2/3}$ at about 5.30 pm and 6.45 am when Ws, T are high: 5.1 m s^{-1}, 35 °C: RH, P are low: 38 %, 101.1 kPa and Ws, T, P are low: 0.2 m s^{-1}, 25 °C, 101.2 kPa; RH is high: 80 % respectively. Local minima are seen in the early morning and late evening at $7.14 \cdot 10^{-16}\,m^{-2/3}$ and $5.46 \cdot 10^{-20}\,m^{-2/3}$ about 4.20 am and 5.30 pm respectively. A greatly fluctuating turbulence pattern is predicted by the PAMELA model during 6.00 am to 6.45 pm, since the model is highly sensitivity to Ws. The results yield less correlation with the measurements at almost all times of day except few a instances

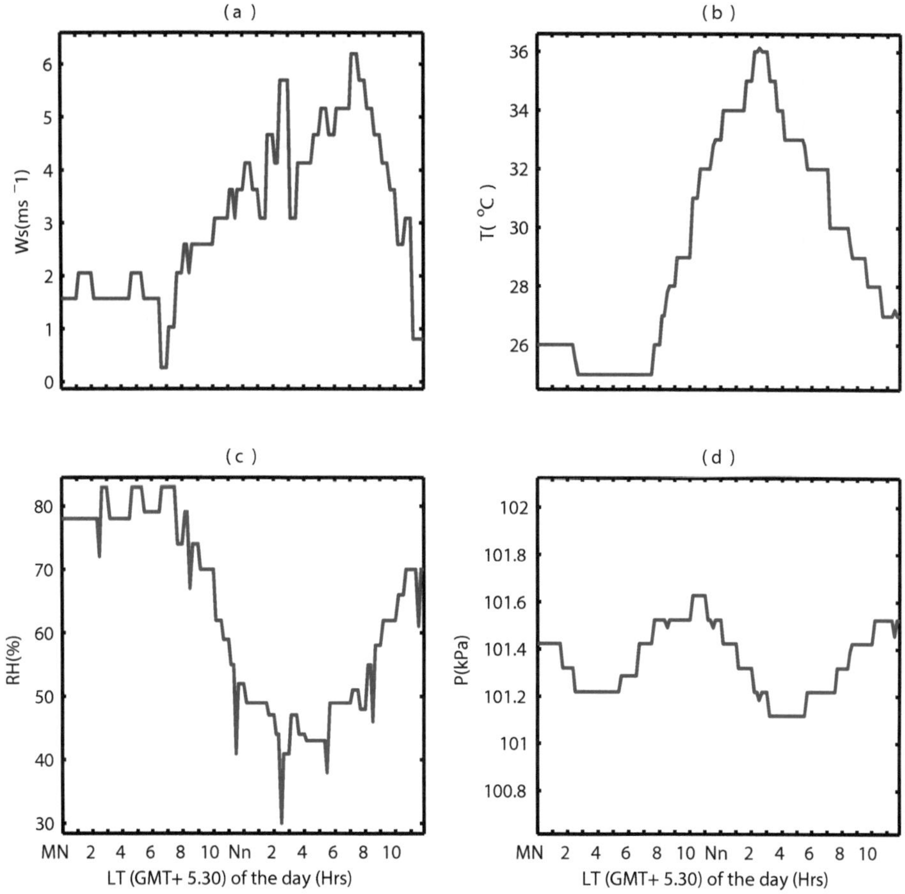

Fig. 3.18: Diurnal time series profile of meteorological parameters: wind speed, temperature, relative humidity and barometric pressure (a–d).

at $3.32 \cdot 10^{-14}\,\mathrm{m}^{-2/3}$, $2.74 \cdot 10^{-14}\,\mathrm{m}^{-2/3}$, $1.80 \cdot 10^{-13}\,\mathrm{m}^{-2/3}$, $1.99 \cdot 10^{-13}\,\mathrm{m}^{-2/3}$, and $4.751 \cdot 10^{-14}\,\mathrm{m}^{-2/3}$ about around 3.00 am, 4.15 am, 11.15 am, 3.10 pm, and 8.00 pm respectively as shown in Fig. 3.19 (a). The maximum deviation is $2.22 \cdot 10^{-10}\,\mathrm{m}^{-2/3}$ and therefore, this model cannot be used to get an accurate prediction of the C_n^2 at our test field. The min/max value of C_n^2 predicted by the Hufnagel–Valley model is $1.65 \cdot 10^{-14}\,\mathrm{m}^{-2/3}$ with SD of $1.9 \cdot 10^{-29}\,\mathrm{m}^{-2/3}$ i.e. this model yields a constant C_n^2 (singularity) throughout the day as in Fig. 3.19 (a) with the maximum deviation of $1.67 \cdot 10^{-13}\,\mathrm{m}^{-2/3}$. This result clearly shows that this model could not keep much dependency on the Ws variations at the 15.25 m altitude and may yield reasonable results as a function of higher altitude. This model shows a somewhat better correlation at $3.02 \cdot 10^{-14}\,\mathrm{m}^{-2/3}$ about 4.00 am in this season. However, this model is unfit for our test field.

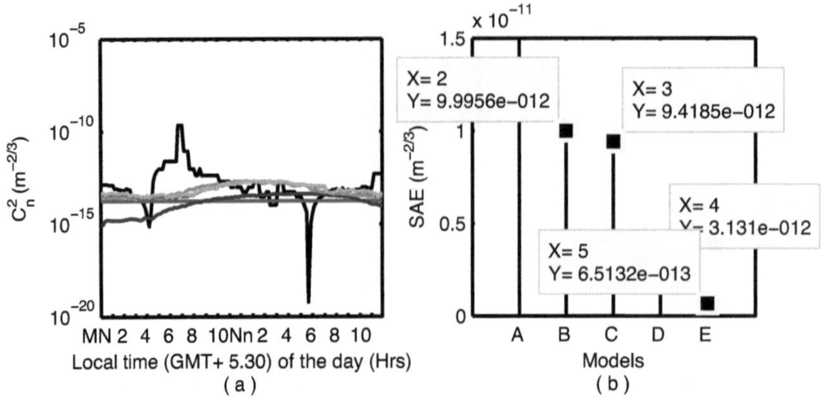

Fig. 3.19: Graphical comparison of time series profile of atmospheric turbulence strength (C_n^2) predicted using selected existing models and proposed models and measured from direct beam transmission system (a). Sum of absolute error value against the particular model (b).

The min/max values of C_n^2 predicted based on beam wandering are $8.69 \cdot 10^{-16}$ $m^{-2/3}$ and $4.29 \cdot 10^{-14}\,m^{-2/3}$ respectively with SD of $1.44 \cdot 10^{-14}\,m^{-2/3}$ at about MN and 2.30 pm to 6.00 pm when the Ws, T are low: 1.5 m s^{-1}, 26 °C; RH, P are high: 78 %, 101.4 kPa and Ws, T are high: 5.9 m s^{-1}, 35 °C; RH, P are low: 44 %, 101.1 kPa respectively. The model predicts a smooth turbulence pattern roughly similar to measurements with a maximum deviation of $4.87 \cdot 10^{-14}\,m^{-2/3}$ as shown in Fig. 3.19 (a). The results meet the measurements at $5.20 \cdot 10^{-14}\,m^{-2/3}$ at about 6.50 pm when Ws, T, P are normal: 4.5 m s^{-1}, 32 °C, 101.2 kPa and RH is low: 50 %. This model is unsuitable at our test field. The min/max values of the polynomial regression model are $2.03 \cdot 10^{-14}\,m^{-2/3}$ and $2.03 \cdot 10^{-13}\,m^{-2/3}$ respectively with SD of $6.32 \cdot 10^{-14}\,m^{-2/3}$ at about MN and 11.00 am to 4.00 pm when Ws, T are low: 1.5 m s^{-1}, 26 °C; RH is high: 78 % and Ws, T keep increasing while RH decreases as in Fig. 3.18 (a–c). This model exhibits a good correlation throughout the day in this season as in Fig. 3.19 (a) except for a few instances at $4.19 \cdot 10^{-14}\,m^{-2/3}$, $7.26 \cdot 10^{-14}\,m^{-2/3}$, $1.05 \cdot 10^{-13}\,m^{-2/3}$, and $3.99 \cdot 10^{-14}\,m^{-2/3}$ at about 1.30 am, 7.45 am, 6.15 pm, and 11.05 pm respectively with the maximum deviation of $5.47 \cdot 10^{-14}\,m^{-2/3}$. The min/max values of the proposed model are $2.35 \cdot 10^{-14}\,m^{-2/3}$ and $1.94 \cdot 10^{-13}\,m^{-2/3}$ respectively with SD of $5.91 \cdot 10^{-14}\,m^{-2/3}$ at about MN and 11.15 am to 3.50 pm when the Ws, T are low: 1.5 m s^{-1}, 26 °C; RH is high: 78 % and Ws, T keep increasing while RH decreases as in Fig. 3.18 (a–c) and Fig. 3.19 (a). The proposed model yields a great correlation with the measurements at all times of day. The computed SAE using the measured and predicted values are shown in Fig. 3.19 (b) against the respective model and prove that the proposed model gives a much lower SAE of $0.000649 \cdot 10^{-9}\,m^{-2/3}$ whereas the other models are $0.7310 \cdot 10^{-9}\,m^{-2/3}$, $0.0107 \cdot 10^{-9}\,m^{-2/3}$, $0.0101 \cdot 10^{-9}\,m^{-2/3}$, and $0.00237 \cdot 10^{-9}\,m^{-2/3}$.

3.6.2.2 Data for 31st May 2013, summer

The weather parameters varied from 0.27 m s^{-1} to 6.16 m s^{-1} with SD of 1.42 m s^{-1} for wind speed, 26 °C to 36 °C with SD of 3 °C for temperature, 39 % to 89 % with SD of 14.54 % for relative humidity and 100.3 kPa to 100.7 kPa with SD of 0.128 kPa for pressure as shown in Fig. 3.20 (a–d). The SD of RH is significantly high while Ws, T, and P are low. The different atmospheric conditions observed on 31st May 2013 (Friday) are mostly cloudy, hazy, scattered cloudy, partial cloudy, and light drizzle. The Ws oscillates randomly around 2.59 m s^{-1} in the morning and 4.11 m s^{-1} in the afternoon.

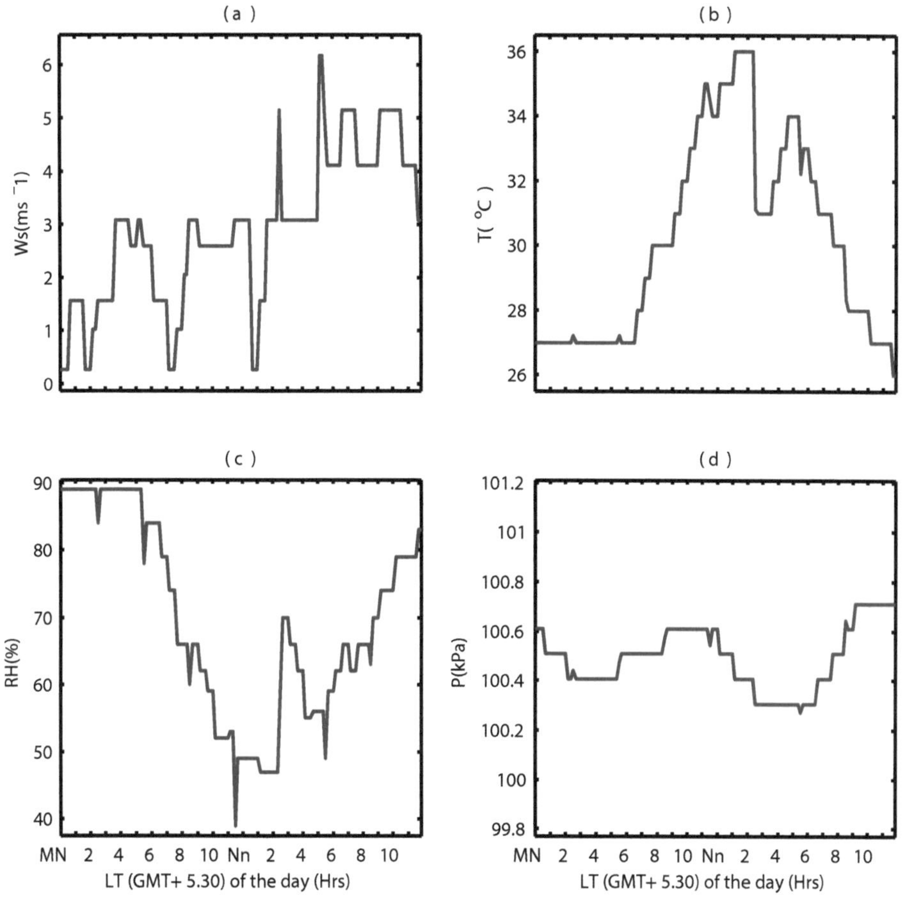

Fig. 3.20: Diurnal time series profile of meteorological parameters: wind speed, temperature, relative humidity and barometric pressure (a–d).

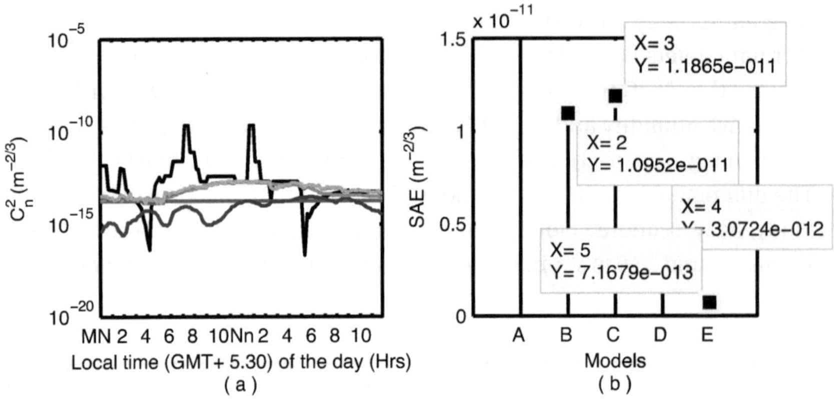

Fig. 3.21: Graphical comparison of time series profile of atmospheric turbulence strength (C_n^2) predicted using selected existing models and proposed models and measured from direct beam transmission system (a). Sum of absolute error value against the particular model (b).

Figure 3.21 (a) shows the experimental time series plot of C_n^2 predicted from local meteorological data collected on 31st May 2013 and measured values. The min/max values of measured C_n^2 are $\approx 1.35 \cdot 10^{-14}\,\text{m}^{-2/3}$ and $1.94 \cdot 10^{-13}\,\text{m}^{-2/3}$ at about MN and 10.00 am to 4.30 pm respectively with SD of $5.61 \cdot 10^{-14}\,\text{m}^{-2/3}$ when Ws, T are low: $1.02\,\text{m}\,\text{s}^{-1}$, 27 °C; RH is high: 89 % and Ws is normal: $3.08\,\text{m}\,\text{s}^{-1}$; T is high: 37 °C; RH is low: 60 % as in Fig. 3.20 (a–c). Sustained and smooth steady variations are seen in the measured turbulence pattern. Comparing the selected models: the min/max values of C_n^2 predicted by the PEMELA model are $2.138 \cdot 10^{-17}\,\text{m}^{-2/3}$ and $2.178 \cdot 10^{-10}\,\text{m}^{-2/3}$ with SD of $4.27 \cdot 10^{-11}\,\text{m}^{-2/3}$ at about 5.05 pm and 7.10 am and 1.13 pm respectively. Local minima existed in the early morning and late evening at $3.90 \cdot 10^{-17}\,\text{m}^{-2/3}$ and $2.13 \cdot 10^{-17}\,\text{m}^{-2/3}$ at about 4.05 am and 5.05 pm respectively. A greatly sporadic C_n^2 pattern is predicted with the maximum deviation of $1.21 \cdot 10^{-13}\,\text{m}^{-2/3}$ since unrealistical fluctuations are seen in the meteorological variations (except pressure). The prediction results give a poor correlation with the measurements in general and meet it randomly at $2.77 \cdot 10^{-14}\,\text{m}^{-2/3}$, $2.37 \cdot 10^{-14}\,\text{m}^{-2/3}$, $1.92 \cdot 10^{-13}\,\text{m}^{-2/3}$, $9.75 \cdot 10^{-14}\,\text{m}^{-2/3}$ and $1.55 \cdot 10^{-13}\,\text{m}^{-2/3}$ at about 2.45 am, 4.30 am, 1.00 pm, 1.55 pm, and 5.00 pm respectively as in Fig. 3.20 (a–d) and Fig. 3.21 (a). A close correlation is seen from 8.45 pm onwards when the Ws is normal: $4\,\text{m}\,\text{s}^{-1}$; T is low: 28 %, RH is high: 79 % and P is high: 100.7 kPa.

The min/max values of the Hufnagel–Valley model are $1.65 \cdot 10^{-14}\,\text{m}^{-2/3}$ with SD of $1.9 \cdot 10^{-29}\,\text{m}^{-2/3}$ i.e. a constant C_n^2 is predicted throughout the day. This model exhibits a reasonable fit with the measurements at $2.59 \cdot 10^{-14}\,\text{m}^{-2/3}$ from MN to 5.05 am. The deflection began from 5.05 am onwards with the maximum value of $1.50 \cdot 10^{-13}\,\text{m}^{-2/3}$. The min/max values of beam wandering model are $2.60 \cdot 10^{-16}\,\text{m}^{-2/3}$ and $2.84 \cdot 10^{-14}\,\text{m}^{-2/3}$ with SD of $8.50 \cdot 10^{-15}\,\text{m}^{-2/3}$ at about MN and 5.15 pm respectively. The prediction results always fall below the measurements with the maximum deviation of $1.46 \cdot 10^{-13}\,m^{-2/3}$. An oscillating pattern of C_n^2 is seen in the morning due to the greatest fluctuations in Ws as shown in Fig. 3.21 (a) and Fig. 3.20 (c). The deviation decreases in the afternoon and becomes close to the results predicted by the Hufnagel–Valley model. The min/max values of the polynomial regression model are $1.74 \cdot 10^{-14}\,m^{-2/3}$ and $1.93 \cdot 10^{-13}\,m^{-2/3}$ with SD of $6 \cdot 10^{-14}\,m^{-2/3}$ respectively at around MN and 10.00 am to 2.00 pm when Ws fluctuates about 2.59 m s^{-1}, T increases: 30 °C to 36 °C and RH decreases: 62 % to 47 % as in Fig. 3.20 (a–c). This model keeps a good track on measurements from MN to 5.15 pm and the deviation began with the maximum deviation of $3.5 \cdot 10^{-14}\,m^{-2/3}$. The min/max values of proposed model are $1.37 \cdot 10^{-14}\,m^{-2/3}$ and $1.87 \cdot 10^{-13}\,m^{-2/3}$ with SD of $5.64 \cdot 10^{-14}\,m^{-2/3}$ at about MN and 10.00 am to 2.10 pm respectively. The proposed model exhibits a greater correlation with the measurements as shown in Fig. 3.21 (a). The SAE results shown in Fig. 3.21 (b) prove that the proposed model exhibits a much lower value of $0.000653 \cdot 10^{-8}\,m^{-2/3}$ whereas other models are $0.134 \cdot 10^{-8}\,m^{-2/3}$, $0.0010 \cdot 10^{-8}\,m^{-2/3}$, $0.0011 \cdot 10^{-8}\,m^{-2/3}$ and $0.00308 \cdot 10^{-9}\,m^{-2/3}$.

3.6.2.3 Data for 14th June 2013, monsoon

The weather parameters varied from 4.11 m s^{-1} to 10.82 m s^{-1} with SD of 2.09 m s^{-1} for Ws, 28 °C to 36 °C with SD of 2.73 °C for temperature, 25 % to 62 % with SD of 9.62 % for relative humidity and 100 kPa to 100.4 kPa with SD of 0.11 kPa for pressure as shown in Fig. 3.22 (a–d). The SD of RH is significantly large while the Ws, T, and P are very low. The different atmospheric conditions observed on 14th June 2013 (Friday) are light drizzle, drizzle and hazy. Greatly uneven Ws variation is observed as in Fig. 3.22 (a).

Figure 3.23 (a) shows the experimental time series plot of C_n^2 predicted from local meteorological data collected on 14th June 2013 and measured values. The min/max values of measured C_n^2 are $\approx 3.63 \cdot 10^{-14}\,m^{-2/3}$ and $1.87 \cdot 10^{-13}\,m^{-2/3}$ at about MN and 11.00 am to 12.30 pm respectively with SD of $4.50 \cdot 10^{-14}\,m^{-2/3}$ when Ws, T, P are low: 5.1 m s^{-1}, 29 °C, 100.1 kPa; RH is high: 62 % and Ws, T, P are high: 10.82 m s^{-1}, 35 °C, 100.34 kPa; RH is low: 39 %. Comparing the selected models: the min/max values of C_n^2 predicted by the PAMELA model are $4.47 \cdot 10^{-18}\,m^{-2/3}$ and $3.21 \cdot 10^{-13}\,m^{-2/3}$ with SD of $8.78 \cdot 10^{-14}\,m^{-2/3}$ respectively at about 4.30 am and 11.50 pm. Local minima are obtained at $4.47 \cdot 10^{-18}\,m^{-2/3}$ and $8.39 \cdot 10^{-18}\,m^{-2/3}$ at about 4.20 am and 4.00 pm respectively. The prediction results fall below the measurements almost all the time and fall above from 6.00 pm to MN with the maximum deviation of $1.49 \cdot 10^{-13}\,m^{-2/3}$. The results keep a reasonable correlation at around 1.30 am since Ws, T are low: 5.1 m s^{-1}, 29 °C and RH, P are high: 62 %, 100.3 kPa. The great fluctuations in Ws yield the unrealistically large sporadic C_n^2 pattern in the daytime as in Fig. 3.22 (a) and Fig. 3.23 (a). The min/max values of the C_n^2 predicted by the Hufnagel–Valley model is $1.658 \cdot 10^{-14}\,m^{-2/3}$ with SD of $1.9 \cdot 10^{-29}\,m^{-2/3}$ i.e. a constant C_n^2 is given by the model throughout the day. The prediction results are away from the measured values

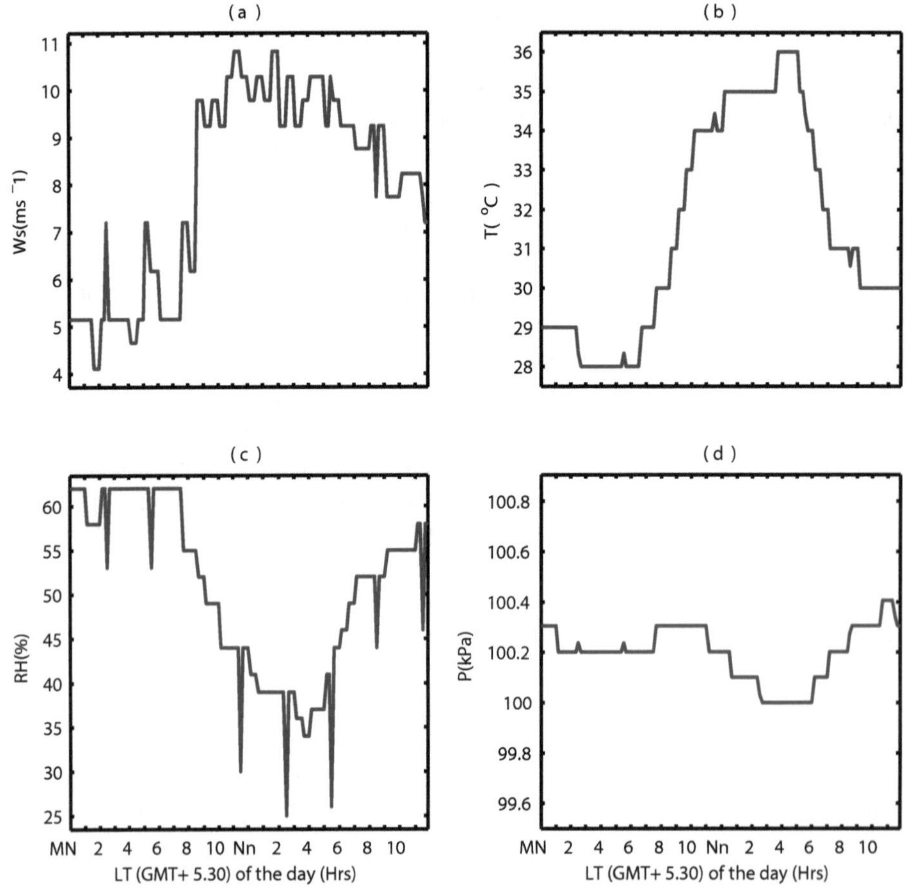

Fig. 3.22: Diurnal time series profile of meteorological parameters: wind speed, temperature, relative humidity and barometric pressure (a–d).

with the maximum deviation of $1.65 \cdot 10^{-14}\,\mathrm{m}^{-2/3}$ as in Fig. 3.23 (a). The min/max values of the C_n^2 predicted by the beam wandering model are $7.36 \cdot 10^{-16}\,\mathrm{m}^{-2/3}$ and $5.4 \cdot 10^{-14}\,\mathrm{m}^{-2/3}$ with SD of $1.66 \cdot 10^{-14}\,\mathrm{m}^{-2/3}$ at about MN and 4.15 pm to 7.00 pm. The results very slowly increase to the maximum from MN to 6.40 pm and decrease to minima. The results fall below the measurements throughout the day with the maximum deviation of $6.52 \cdot 10^{-14}\,\mathrm{m}^{-2/3}$.

The correlation between prediction and measurement is very low as shown in Fig. 3.23 (a). The min/max values of C_n^2 predicted by the polynomial regression model are $3.31 \cdot 10^{-14}\,\mathrm{m}^{-2/3}$ and $2 \cdot 10^{-13}\,\mathrm{m}^{-2/3}$ with SD of $5.51 \cdot 10^{-14}\,\mathrm{m}^{-2/3}$ at about MN and 11.00 am to 4.00 pm respectively when the Ws, T are low: 5.14 m s^{-1}, 29 °C; RH is high: 62 %; Ws, T are high: 10.82 m s^{-1}, 36 °C; RH is low: 38 %. The prediction results fail to maintain a good correlation with the measurements throughout the day

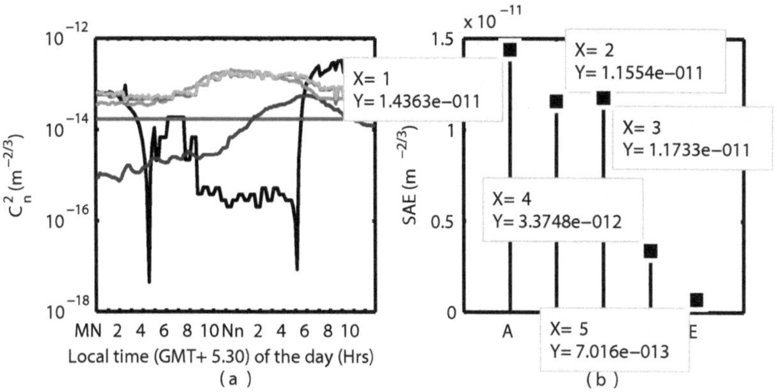

Fig. 3.23: Graphical comparison of time series profile of atmospheric turbulence strength (C_n^2) predicted using selected existing models and proposed models and measured from direct beam transmission system (a). Sum of absolute error value against the particular model (b).

unlike its behavior in the previous seasons. The results appear above the measurements from 7.00 am to 2.00 pm with the maximum deviation of $8.7 \cdot 10^{-14}$ $m^{-2/3}$. However, the results fall below the measurements with greater deviations during MN to 6.00 am and 3.00 pm to MN as in Fig. 3.23 (a). The min/max values of the proposed model are $3.18 \cdot 10^{-14}$ $m^{-2/3}$ and $1.7 \cdot 10^{-13}$ $m^{-2/3}$ with SD of $4.46 \cdot 10^{-14}$ $m^{-2/3}$ at about MN and 11.00 am to 3.30 pm respectively. The proposed model exhibits a greater correlation with the measurements as shown in Fig. 3.23 (a). The SAE results shown in Fig. 3.23 (b) prove that the proposed model exhibits a much lower value of $0.00708 \cdot 10^{-10}$ $m^{-2/3}$ whereas the other models give $0.143 \cdot 10^{-10}$ $m^{-2/3}$, $0.1156 \cdot 10^{-10}$ $m^{-2/3}$, $0.1173 \cdot 10^{-10}$ $m^{-2/3}$, and $0.0336 \cdot 10^{-10}$ $m^{-2/3}$.

3.6.2.4 Data for 21st November 2013, rainy

The weather parameters varied from 0.25 m s^{-1} to 4.11 m s^{-1} with SD of 0.90 m s^{-1} for wind speed, 25.2 °C to 31.36 °C with SD of 2.26 °C for temperature, 55 % to 94 % with SD of 11.67 % for relative humidity and 100.5 kPa to 101 kPa with SD of 0.146 kPa for pressure as shown in Fig. 3.24 (a–d). The SD of T and RH are high while Ws and P are low. The different atmospheric conditions observed on 21st November 2013 (Thursday) are hazy, scattered cloudy, fogy and misty.

Figure 3.25 (a) shows the experimental time series plot of C_n^2 predicted from local meteorological data collected on 21st November 2013 and measured values. The min/-max values of measured C_n^2 are $\approx 3.43 \cdot 10^{-14}$ $m^{-2/3}$ and $1.40 \cdot 10^{-13}$ $m^{-2/3}$ at about MN and 11.00 am to 4.00 pm respectively with SD of $3.20 \cdot 10^{-14}$ $m^{-2/3}$. Comparing the selected models: the min/max values of C_n^2 predicted by the PAMELA model are $4.82 \cdot 10^{-17}$ $m^{-2/3}$ and $2.60 \cdot 10^{-10}$ $m^{-2/3}$ with SD of $1.05 \cdot 10^{-10}$ $m^{-2/3}$ at about 2.10 am and 5.10 am to 10.05 am respectively. The prediction results are highly sporadic as is the

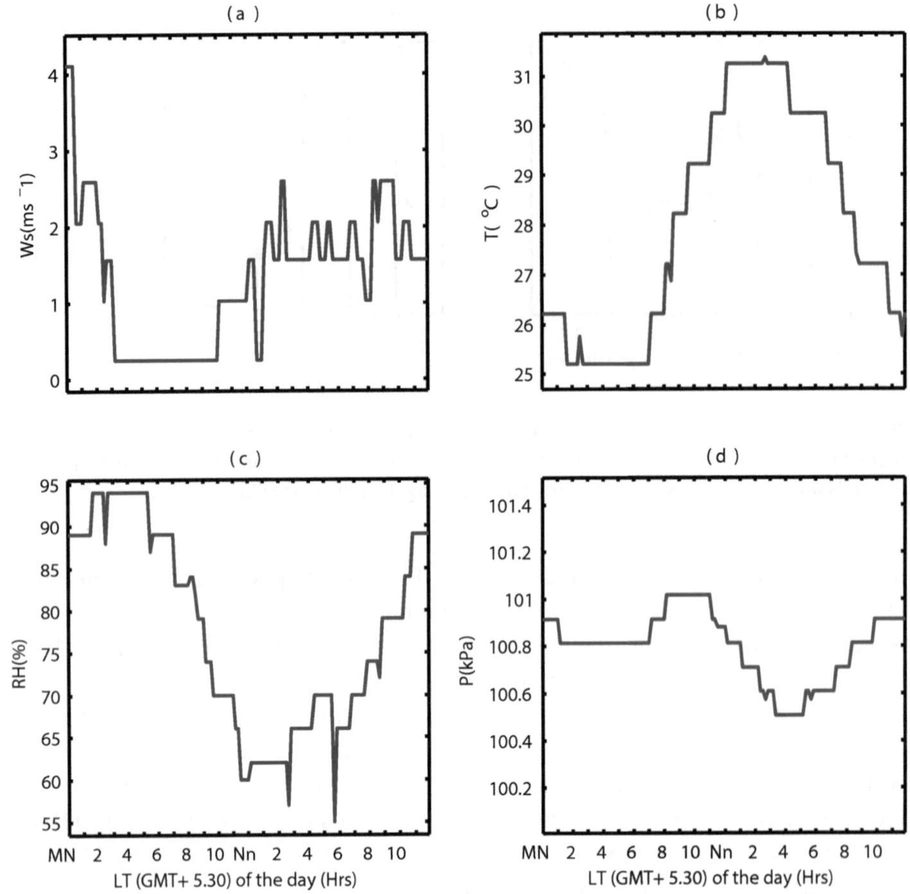

Fig. 3.24: Diurnal time series profile of meteorological parameters: wind speed, temperature, relative humidity and barometric pressure (a–d).

Ws in Fig. 3.24 (a). Most of the time, the results fall above the measurements as in Fig. 3.25 (a) with the maximum deviation of $2.56 \cdot 10^{-10}\,\mathrm{m}^{-2/3}$. This model exhibits two local minima at $4.82 \cdot 10^{-17}\,\mathrm{m}^{-2/3}$ and $4.68 \cdot 10^{-16}\,\mathrm{m}^{-2/3}$ at about 2.05 am and 5.55 pm respectively with a reasonable correlation during 8.00 pm to MN since Ws is calm: $0.27\,\mathrm{m\,s^{-1}}$, T is low: 26.21 °C; RH is high: 90 % and P is high: 100.9 kPa.

The min/max values of the C_n^2 predicted by the Hufnagel–Valley model is $1.658 \cdot 10^{-14}\,\mathrm{m}^{-2/3}$ with SD of $1.9 \cdot 10^{-29}\,\mathrm{m}^{-2/3}$ i.e. a constant C_n^2 is given by the model throughout the day and appears above the measurements. The prediction results are not correlated with measurements with the maximum deviation of $1.23 \cdot 10^{-13}\,\mathrm{m}^{-2/3}$ as in Fig. 3.25 (a). The min/max values of C_n^2 predicted by the beam wandering model are $6.13 \cdot 10^{-16}\,\mathrm{m}^{-2/3}$ and $1.94 \cdot 10^{-14}\,\mathrm{m}^{-2/3}$ with SD of $5.60 \cdot 10^{-15}\,\mathrm{m}^{-2/3}$ at about 4.10 am and 5.05 pm respectively. The prediction results appear below the measure-

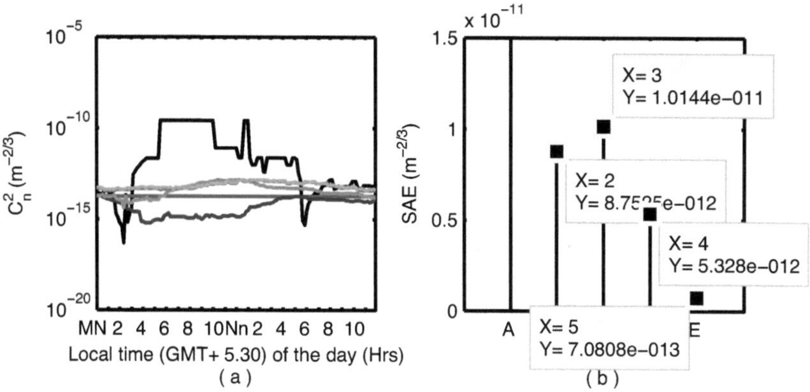

Fig. 3.25: Graphical comparison of time series profile of atmospheric turbulence strength (C_n^2) predicted using selected existing models and proposed models and measured from direct beam transmission system (a). Sum of absolute error value against the particular model (b).

ments with the maximum deviation of $5.32 \cdot 10^{-14}\,m^{-2/3}$. The deviation decreases from 1.00 pm onwards and becomes close to the response of the Hufnagel–Valley model. However, the prediction results are irrelevant to the measurements throughout the diurnal period. The min/max values of C_n^2 predicted by the polynomial regression model are $3.98 \cdot 10^{-15}\,m^{-2/3}$ and $1.42 \cdot 10^{-13}\,m^{-2/3}$ with SD of $4.04 \cdot 10^{-14}\,m^{-2/3}$ at about MN and 10.00 am to 4.00 pm respectively. This model exhibits a reasonable correlation with the measurements in the morning session from 7.00 am to 1.00 pm and afterwards, the deviation exists with the maximum value of $4.62 \cdot 10^{-14}\,m^{-2/3}$ as predicted in the previous season. The min/max values of the proposed model are $3.35 \cdot 10^{-14}\,m^{-2/3}$ and $1.31 \cdot 10^{-13}\,m^{-2/3}$ with SD of $3.22 \cdot 10^{-14}\,m^{-2/3}$ at about MN and 10.30 am to 4.30 pm respectively. The proposed model exhibits a greater correlation with the measurements as shown in Fig. 3.25 (a). The SAE results shown in Fig. 3.25 (b) shows that the proposed model exhibits a much lower value of $0.0000743 \cdot 10^{-8}\,m^{-2/3}$ whereas the other models give $0.8137 \cdot 10^{-8}\,m^{-2/3}$, $0.000878 \cdot 10^{-8}\,m^{-2/3}$, $0.00101 \cdot 10^{-8}\,m^{-2/3}$ and $0.000535 \cdot 10^{-8}\,m^{-2/3}$.

3.6.2.5 Data for 25th December 2013, winter

The weather parameters varied from 2.05 m s^{-1} to 7.19 m s^{-1} with SD of 1.30 m s^{-1} for wind speed, 22.18 °C to 30.24 °C with SD of 2.59 °C for temperature, 40 % to 88 % with SD of 11.28 % for relative humidity and 101.3 kPa to 101.8 kPa with SD of 0.11 kPa for pressure as shown in Fig. 3.26 (a–d). The SD of T and RH are high while Ws and P are low. The different atmospheric conditions observed on 25th December 2013 (Wednesday) are hazy, partially cloudy, misty and scattered cloudy.

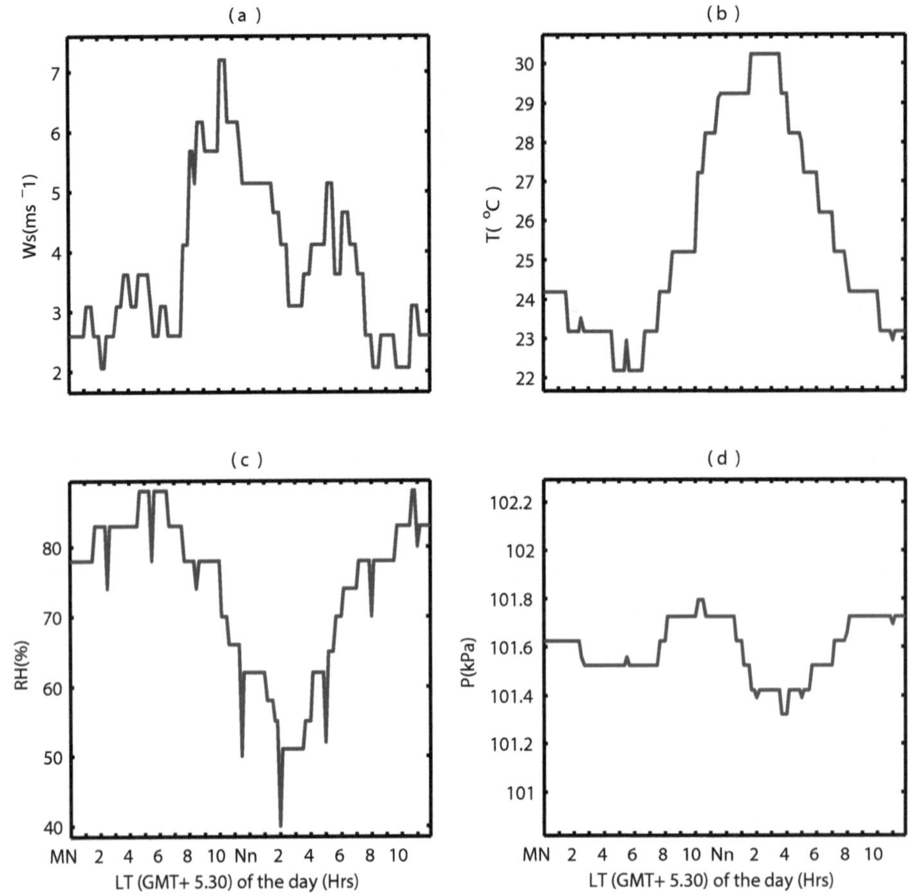

Fig. 3.26: Diurnal time series profile of meteorological parameters: wind speed, temperature, relative humidity and barometric pressure (a–d).

Figure 3.27 (a) shows the experimental time series plot of C_n^2 predicted from local meteorological data collected on 25th December 2013 and measured values. The min/max values of measured C_n^2 are $\approx 1.13 \cdot 10^{-14}\,m^{-2/3}$ and $8.34 \cdot 10^{-14}\,m^{-2/3}$ respectively at about MN and 9.00 am to 4.00 pm with SD of $2.06 \cdot 10^{-14}\,m^{-2/3}$ when Ws, T are low: 2.5 m s^{-1}, 24.19 °C; RH, P are high: 78 %, 101.6 kPa and Ws, T are high: 6.19 m s^{-1}, 30.24 °C; RH, P are low: 50 %, 101.2 kPa. The larger fluctuations in the meteorological values generate unrealistically large sporadic C_n^2 values. Comparing the selected models: the min/max values of C_n^2 predicted by the PAMELA model are $3.68 \cdot 10^{-17}\,m^{-2/3}$ and $4.17 \cdot 10^{-13}\,m^{-2/3}$ with SD of $1.03 \cdot 10^{-13}\,m^{-2/3}$ at about 3.15 am and around 7.00 am respectively. The prediction results did not match with the measurements and the maximum deviation is $4.00 \cdot 10^{-13}\,m^{-2/3}$. Two local minima exist at $3.68 \cdot 10^{-17}\,m^{-2/3}$ and $4.74 \cdot 10^{-17}\,m^{-2/3}$ at about 3.15 am and 6.00 pm respectively as in Fig. 3.27 (a). The prediction pattern approximately oscillates about the measurements. The C_n^2 pattern clearly exhibits the great dependency of the PAMELA model on Ws. The min/max values of the C_n^2 predicted by the Hufnagel–Valley model is $1.65 \cdot 10^{-14}\,m^{-2/3}$ with SD of $1.9 \cdot 10^{-29}\,m^{-2/3}$ i.e. a flat behavior with a constant C_n^2 is given by the model throughout the day and appears below the measurements. The prediction results are not correlated with measurements with the maximum deviation of $5.35 \cdot 10^{-14}\,m^{-2/3}$ as in Fig. 3.27 (a).

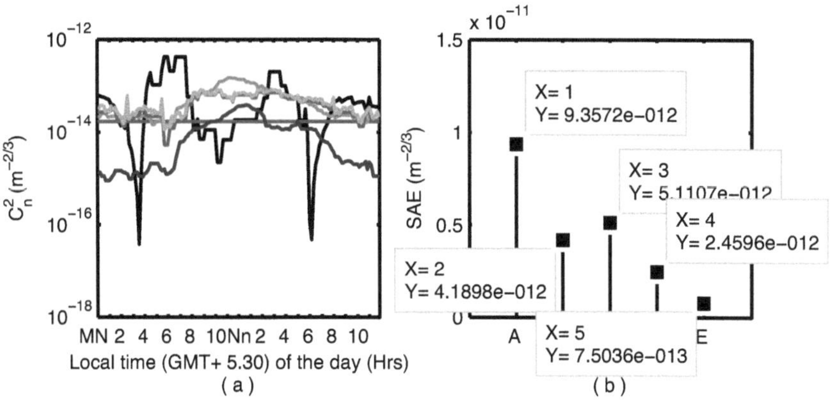

Fig. 3.27: Graphical comparison of time series profile of atmospheric turbulence strength (C_n^2) predicted using selected existing models and proposed models and measured from direct beam transmission system (a). Sum of absolute error value against the particular model (b).

The min/max values of C_n^2 predicted by the beam wandering model are $8.69 \cdot 10^{-16}$ and $3.78 \cdot 10^{-14}\,m^{-2/3}$ with SD of $1.03 \cdot 10^{-14}\,m^{-2/3}$ at about MN and Nn respectively. A sustained and steady increasing and decreasing C_n^2 pattern is seen throughout the diurnal period as in Fig. 3.27 (a). The result exhibits approximately the same pattern as the measurements with a maximum deviation of $3.14 \cdot 10^{-14}\,m^{-2/3}$. However, the results fall below the measurements throughout the day. The min/max value of C_n^2 predicted by the polynomial regression model is $1.38 \cdot 10^{-15}\,m^{-2/3}$ and $1.49 \cdot 10^{-13}\,m^{-2/3}$ at about MN and 11.00 am to 1.30 pm respectively. The prediction results yield a reasonable correlation with the measurements from MN to 9.00 am and 3.00 pm to MN. A greatest deviation existed around Nn with the maximum deviation of $8.43 \cdot 10^{-14}\,m^{-2/3}$. The results absolutely match the measurements only for some instances at $2.19 \cdot 10^{-14}\,m^{-2/3}$, $1.29 \cdot 10^{-14}\,m^{-2/3}$, $7.49 \cdot 10^{-14}\,m^{-2/3}$, $5.83 \cdot 10^{-14}\,m^{-2/3}$, and $4.86 \cdot 10^{-14}\,m^{-2/3}$ at about 2.00 am, 6.05 am, 9.05 am, 3.45 pm, and 6.05 pm respectively as in Fig. 3.27 (a). The min/max values of the proposed model are $5.11 \cdot 10^{-15}\,m^{-2/3}$ and $7.58 \cdot 10^{-13}\,m^{-2/3}$ with SD of $2.05 \cdot 10^{-14}\,m^{-2/3}$ at about MN and 10.00 am to 3.45 pm respectively. The proposed model exhibits a greater correlation with the measurements as shown in Fig. 3.27 (a). The SAE results shown in Fig. 3.27 (b) evidence that the proposed model exhibits a much lower value of $0.07503 \cdot 10^{-11}\,m^{-2/3}$ whereas the other models give $0.935 \cdot 10^{-11}\,m^{-2/3}$, $0.418 \cdot 10^{-11}\,m^{-2/3}$, $0.511 \cdot 10^{-11}\,m^{-2/3}$, and $0.245 \cdot 10^{-11}\,m^{-2/3}$.

3.7 Summary

The significance of model design for optical attenuation and turbulence strength are explained along with the measurement technique for visibility and attenuation. The review results of background and related works are presented. The establishment of experimental setups (transmitter and receiver) built for model development and practical validation are explained. The models selected for comparative analysis are discussed. The formulation of mathematical model are detailed and based on the results of the coefficient of determination (R^2) obtained from ANOVA tools, the Eq. (3.19): $R^2 = 98.76\,\%$ and Eq. (3.24): $R^2 = 98.93\,\%$ for optical attenuation and turbulence strength respectively are finalized. Average percentage deviation of 1.36 % and 3.86 % are obtained for optical attenuation and turbulence strength respectively from the confirmatory test. The seasonal average RMSE of 0.043 dB/km, 0.040 dB/km, 0.053 dB/km, 0.049 dB/km, and 0.051 dB/km and SAE of $0.00064 \cdot 10^{-9}\,m^{-2/3}$, $0.000653 \cdot 10^{-8}\,m^{-2/3}$, $0.00708 \cdot 10^{-10}\,m^{-2/3}$, $0.0000743 \cdot 10^{-8}\,m^{-2/3}$, and $0.075 \cdot 10^{-11}\,m^{-2/3}$ are achieved for five different seasons: monsoon, rainy, winter, presummer and summer respectively in a one year period. These results evidence the suitability and feasibility of the proposed models to have more accurate prediction results at our test field throughout the duration of experimentations (Bazil Raj et al. 2015a, b, with permission of De Gruyter and OSA).

4 Mitigation of beam wandering due to atmospheric turbulence and prediction of control quality using intelligent decision making tools

In a Free Space Optical Link (FSOL), atmospheric turbulence causes fluctuations in both intensity and phase of the received beam and impairs link performance. Beam motion is one of the main causes for major power loss. This chapter presents an investigation of the performance of two types of controller designed for aiming a laser beam at a particular spot under dynamic disturbances. Multiple experiment observability nonlinear input-output data mapping is used as the principal component for the controllers' design. The first design is based on the Taguchi method while the second is the Artificial Neural Network (ANN) method. These controllers process the beam location information from a static linear map of a 2D plane: Optoelectronic Position Detector (OPD) as observer, and then generate the necessary outputs to steer the beam with a micro-electromechanical mirror: Fast Steering Mirror (FSM). The beam centroid is computed using a Mono-Pulse Algorithm (MPA). Evidence of suitability and effectiveness of the proposed controllers are comprehensively assessed and quantitatively measured in terms of coefficients of correlation, correction speed, control exactness, centroid displacement and stability of the receiver signal through the experimental results from the FSO link setup established for the horizontal range of 0.5 km at an altitude of 15.25 m. The test field type is open flat terrain, grass and a few isolated obstacles.

4.1 Introduction

A Free Space Optics Link (FSOL), also known as an optical wireless link, is an effective high bandwidth communication technology serving commercial point-to-point links in terrestrial last mile applications and in infrared indoor LANs (Majumdar & Ricklin 2008; Hemmati 2009; Yuksel et al. 2009). The FSOL has several attractive characteristics like license-free band of operation, dense spatial reuse, low power usage per bit, relatively high bandwidth, intrinsic immunity to Electro-Magnetic Interference (EMI), increased security, flexibility, mobile battlefield networks and easy-to-install (Dat et al. 2011; Chow et al. 2013; Arnon 2003; Kumar & Rana 2013). Further, the laser beams are used to transmit energy in a wide range of applications, from laser cutting to meteorology and from laser surgery to communications (Perez-Arancibia et al. 2012). The FSOL can be made less susceptible to unwanted detection than radio-frequency communication because it is possible to concentrate an optical transmission in a narrow beam aimed toward the intended recipient (Wang et al. 2002). In the former case FSOL is mainly considered for last-mile applications in urban areas where the deployment of fiber optic links can be much more expensive and take much more time than the installation of a pair of terminals for FSO communication system. In recent years, FSO applications have generated much interest in the telecommunications community for both ground and space wireless links at high data rates. The FSO technology can be successfully used in various application scenarios which include space communications e.g. inter-satellite and deep space (Konesky 2006) and terrestrial communications e.g. enterprise connectivity and backup links (Ciaramella et al. 2009). Optical communications through the atmosphere are important to commercial and defense applications (Leeb 2000). The FSOL offer potentially huge bandwidth and very high speed that make them an extremely attractive means of meeting the ever increasing demand for broadband traffic, mostly driven by internet access, IP-TV, VoIP, YouTube and HDTV broadcasting services (Ciaramella et al. 2009; Djordjevic 2011). In addition to high bandwidth traffic requirements, the future optical networks should allow the interoperability of radio frequency, fiber optic and FSO technologies (Djordjevic 2011; Wu et al. 2010). The jump from microwaves to light waves means a reduction in beamwidth by orders of magnitude, even if we use transmitter antennas of much smaller diameter. The reduced beamwidth does not only imply increased intensity at the receiver site but also reduces the cross talk between closely operating links and reduces the chances for eavesdropping (Leeb 2000).

However, with FSOL turbulence affects the signal as it propagates through the atmosphere (Si & Zhang 2013). Various applications typically required that the beam is aimed at a target with a maximum possible light intensity level, or above a specified threshold value. Traveling through free space, the atmosphere or other transmissible media disturbs the quality of the optical signal (Perez-Arancibia et al. 2012). Atmospheric turbulence is caused by small fluctuations in temperature, relative humidity, wind speed and pressure as random changes in the atmospheric refractive index structure function (C_n^2). At the receiver focal plane, the variations in C_n^2, vibrations at the transmitter/receiver, building sway and earthquakes cause absorption, scattering, small fluctuations of the received optical intensity (scintillation) (Yuksel et al. 2009; Fidler et al. 2010), beam wandering, wavefront aberration, and polarization drift in the temporal and spatial domain which degrades the overall performance of the optical link (Arnon 2003; Si & Zhang 2013; Kashani et al. 2013). The desired covertness could be easily broken due to beam outage (Wang et al. 2002) and it leads to low Signal to Noise Ratio (SNR) and high Bit Error Rate (BER) (Liu et al. 2014; Navidpour et al. 2004). The major limitations of FSO are Line of Sight (LoS) maintenance for continuous data flow (Yuksel et al. 2009; Fidler et al. 2010) and wavefront correction to significantly decrease the BER (Majumdar & Ricklin 2008; Hemmati 2009; Kumar & Rana 2013; Toyoshima et al. 2001). Therefore, a micro-mechanical auto-tracking, Pointing Acquisition and Tracking (PAT) system with beam steering controller design (Zhen et al. 2009) to mitigate beam wandering (Yuksel et al. 2009; Kashani et al. 2013; Lee & Ortiz 2003) and an adaptive optics system to compensate the wavefront aberration or phase fluctuations (Konesky 2006; Wu et al. 2010; Toyoshima et al. 2001) becomes significant to improving link reliability (Efe et al. 2003) and data rate (Majumdar & Ricklin 2008; Hemmati 2009; Ciaramella et al. 2009; Lu et al. 2013).

Beam wandering and deep signal fading (wavefront distortions) represent unique and significantly more challenging problems (Si & Zhang 2013), which cannot be resolved using conventional data coding techniques (Ficocelli & Amara 2012). A major incentive for the incorporation of adaptive optics technology (beam steering and wavefront corrections) into FSOL is the active prevention of long-term data loss (Majumdar & Ricklin 2008, Konesky 2006; Wu et al. 2010; McGuire et al. 2002). The first and most important approach in FSO is beam steering (Perez-Arancibia et al. 2012; Bazil Raj et al. 2012) which mitigates beam wandering (temporal distortions) (Yuksel et al. 2009; Ciaramella et al. 2009; Leeb 2000; Fidler et al. 2010; Yang et al. 2002) and is the main contribution in this chapter. The complexity of controller design increases, since most of the process are inherently nonlinear (Paterson et al. 2000) thereby making the control of a nonlinear system in a closed loop nontrivial (Shen et al. 2007; Wood et al. 2005). The Artificial Neural Network (ANN) is the most popular intelligent tool due to its high ability to learn from sample data and to yield solutions to the control applications (Hagan & Demuth, 1999; Xie et al. 2009). Beam steering control is attempted using (i) Taguchi's (analysis of variance – ANOVA) method, the Response Surface Model (RSM) and (ii) ANN, the Neural-controller. In this work, both controllers are implemented in MATLAB for generating the control signals in real-time for steering the beam.

The rest of the chapter is organized as follows: Section 4.2 presents the background and related works, Section 4.3 describes the construction of the experimental setup and estimation of beam centroid displacement, Section 4.4 discusses the equipment's steady state response and calibration results, Sections 4.5 and 4.6 explain the development of the RSM and the neural-controller respectively, Section 4.7 discusses the experimental results and data analysis, and Section 4.8 summarizes the chapter.

4.2 Background and related works

An overview of closely related works is given in this section.

Nestor O Perez-Arancibia et al. (2012) developed an indoor beam stabilization system with an observer-based feedback controller. Beam fluctuation is introduced by an optical shaker and corrected by a Fast Steering Mirror (FSM) using the extended Kalman filter developed in the PC. The experimental results are presented.

Ciaramella et al. (2009) present a dedicated electronic control unit that effectively tracked signal beam wandering caused by atmospheric turbulence and mechanical vibrations in a wireless optical data link of 212 m between two buildings in Pisa, Italy. The tracking beam is coupled with the Single Mode Fiber (SMF).

Hanling Wu et al. (2010) describe the disturbances (spatial coherence degradation) of the atmospheric turbulence effects in coupling the laser beam with the SMF.

The incorporation of adaptive optics to mitigate these effects is addressed. Coupling efficiency is improved using the Zernike model adaptive optics and the results are numerically evaluated.

Franz Fidler et al. (2010) present a review of technology, theoretical studies, and experimental trials for FSOL from/to high altitude platforms. The laser beam PAT with low divergence to transmit data at multi-gigabits per second is described. The outage probability of the FSOL system is discussed. The fading effects are measured and analyzed.

Wei Liu et al. (2014) present the significance of beam pointing in FSO communication in terms of BER, pointing error probability density function and intensity displacement on focal plane. The experiment is conducted in an indoor laboratory setup consisting of a turbulence simulation box. The FSM is used to correct the displacement error using the parallel controller developed in a Field Programmable Gate Array (FPGA). The minimum BER of 10^{-9} is achieved.

Morio Toyoshima et al. (2001) address the deformation of received beam mutual alignment error due to temperature variations. The most influencing factor of 3rd order coma aberration on the mutual alignment error is analyzed and tip/tilt tracking control is used. The experimental results with different optical source are described.

Lee and Ortiz (2003) discuss the use of inertial sensors to facilitate deep space optical communications. The principle concept and algorithm for using linear accelerometer along with the simulation results are given. A simple angular position algorithm is described.

Yafei Lu et al. (2013) present the necessity of high bandwidth, response speed and stable precision of FSM. The problems associated with the design of closed loop bandwidth for FSM and controller design are described. The integrated mechanical control simulation method is used to simulate structural response. The relation between FSM elastic support design and bandwidth is illustrated. The geometrical structures of beam steering with FSM are described.

Tsung-Yi Yang et al. (2002) present the adaptive alignment scheme for optoelectronic systems interconnected by an FSOL. The method of using a quadrant detector to detect the alignment error in six degrees of freedom is described. A novel control algorithm in closed loop is presented to eliminate the alignment errors. The experimental results are described.

Jing-Chung Shen et al. (2007) discuss the precision tracking control of piezoelectric actuators. The Prandtl–Ishlinskii (PI) model and sliding-mode controller are used to reduce the hysteresis nonlinearity. The tracking resolutions for multifrequency nonstationary dynamic motion are presented.

Wen-Fang Xie et al. (2009) propose a neural network based adaptive controller for piezoelectric actuators with unknown hysteresis. A neural network based dynamic pre-inversion compensator is designed to cancel out the effect of the hysteresis.

4.3 FSO link – optoelectronic assembly and setup description

The FSO link experimental setup (transmitter and receiver) shown in Fig. 4.1 is developed with the necessary optoelectronic components for a distance of 0.5 km at an altitude of 15.25 m. The schematic diagram of the setup shown in Fig. 4.2 consists of optical source, optical shaker, disturbance generator, Digital to Analog converter (D/A) and transmission optics at the transmitter and telescope, FSM, variable beam splitter, Optoelectronic Position Detector (OPD), Analog to Digital converter (A/D), RSM and neural-controller at the receiver. The FSM consists of three terminal piezoelectric actuators based on a moving platform on which the beam steering mirror is mounted. In the experimental setup, the devices are coordinated on vibration damped optical breadboards. The optical beam that comes from the source falls on the optical shaker and is transmitted to the receiver through the atmospheric turbulent optical channel. The transmission optics are used to increase the beam diameter from 3 mm to 9 mm. The arbitrary disturbance sequences are generated by PC1 and corresponding analog voltage is given to the optical shaker on which a Pure Reflection Mirror (PRM) is mounted to act as the second possible source of jitter. The telescope captures all the optical power and reflects it to fall on the FSM. The incident laser beam of the FSM gets reflected to the variable beam splitter that splits the incident into two beams: reflected beam and transmitted beam.

The reflected beam is made to fall on the photodiode to generate a voltage related by a linear function with negative slope to intensity of light. Since the slope of the linear relation is negative, when the light intensity increases the photodiode output voltage decreases. Thus, if no light is incident on the photodiode, the voltage output

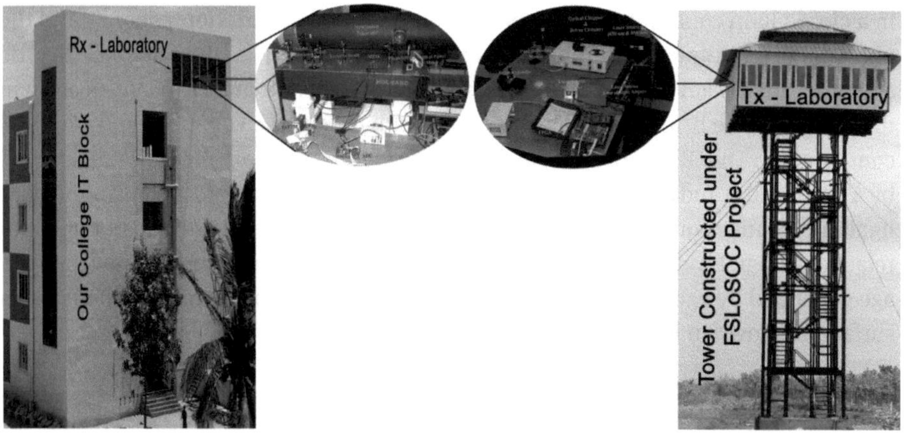

Fig. 4.1: Photograph of Laser Communication Laboratory (LCL) facility: FSOL receiver (left) and transmitter (right) laboratories developed at information technology block and tower constructed for this work respectively.

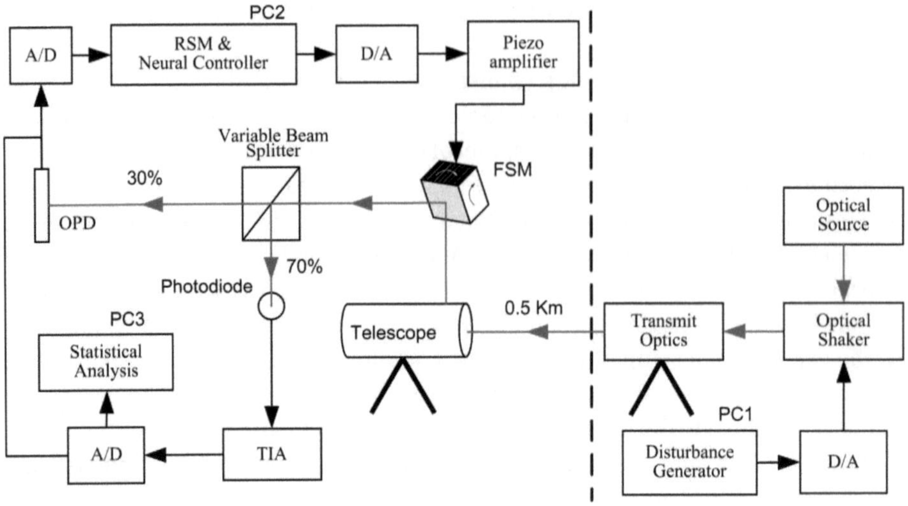

Fig. 4.2: Schematic of FSOL experimental setup for 0.5 km optical link. Left-hand side: receiver with beam steering optoelectronic assembly and right-hand side: transmitter.

is a positive number greater than zero. Conversely, if the amount of light illuminating the photodiode is too large, the sensor saturates and the voltage output is equal to zero (Perez-Arancibia et al. 2012).

The transmitted beam (passed through the beam splitter) falls on the OPD. The outputs of OPD are applied to A/D and the error data are given to the RSM and neural-controller developed in PC2. The RSM and neural-controller response data are converted to an analog signal using D/A and given to the FSM through the piezo-amplifiers in a closed loop control configuration. The OPD outputs, controller response and the photodiode outputs are continuously monitored using PC3. The OPD and piezoelectric actuators are the key devices that facilitate efficient fast and precise beam steering in the FSO link. The beam displacement on the OPD data is the key input and acquired continuously. The OPD consists of four separate identical silicon photodiodes denoted by A, B, C, and D and arranged in quadrant geometry. These detectors convert incident light into relative currents I_A, I_B, I_C, and I_D and then the currents are transformed into relative voltage levels V_A, V_B, V_C, and V_D using the I–V converter circuit. The voltage generated by each quadrant is proportional to the optical energy illuminating its surface as shown in Fig. 4.3.

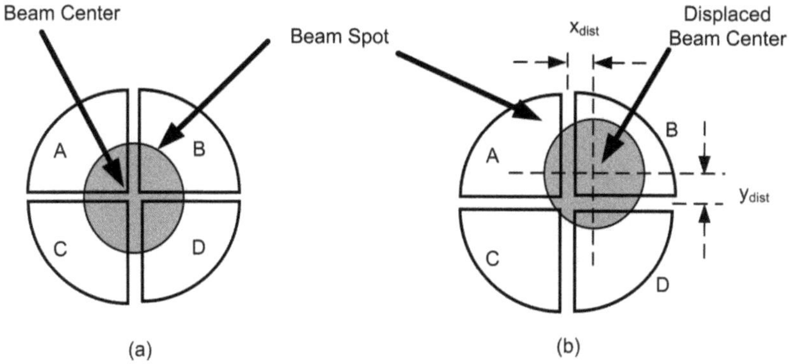

Fig. 4.3: Beam spot of 4 mm diameter (a) centered (0, 0) mm and (b) displaced (X_{dist}, Y_{dist}) mm beam spot on the optoelectronic position detector surface.

The outputs of the OPD are connected to the Mono-Pulse Arithmetic Circuit (MPAC) where the beam spot spatial displacement errors V_{Ex} and V_{Ey} along the x- and y-channels (2D Plane) are measured as relative output voltage changes (Fidler et al. 2010; Yang et al. 2002) by

$$\begin{aligned} V_{Ex} &= \{(V_A + V_C) - (V_B + V_D)\}\,, \\ V_{Ey} &= \{(V_A + V_B) - (V_C + V_D)\}\,. \end{aligned} \tag{4.1}$$

The MPAC is constructed using operational amplifiers for computing the addition and subtraction of signal as given in Eq. (4.1). The reference signal V_{Ref} is measured by the algebraic sum of signals from all the quadrants of OPD (Fidler et al. 2010; Yang et al. 2002) as in Eq. (4.2).

$$V_{Ref} = (V_A + V_B + V_C + V_D)\,. \tag{4.2}$$

The analog output signals V_{Ex}, V_{Ey}, and V_{Ref} vary from −10 V to +10 V and 0 V to +10 V respectively. These signals are applied to the 16 bit data acquisition system (Advantech ADAM 5000 series hardware). This module consists of eight analog inputs, four analog output channels and an RS232 communication port. The RSM and neural-controller algorithm are implemented in real time using MATLAB in PC2. The important statistical values related to the beam displacement on the OPD are: azimuth and elevation distance (mm); radial distance, which is a line connecting the center of the OPD and beam centroid (mm); and beam wandering angle on OPD plane (μrad) and are estimated using Eqs. (4.3)–(4.5).

$$x_{dist} = -2\left(\frac{V_{Ex}}{V_{Ref}}\right) \quad \text{and} \quad y_{dist} = 2\left(\frac{V_{Ey}}{V_{Ref}}\right); \tag{4.3}$$

$$\gamma = \sqrt{x_{dist}^2 + y_{dist}^2}\,; \tag{4.4}$$

$$\vartheta = \begin{cases} 0 & \text{if } V_{Ex} = V_{Ey} = 0, \\ 0 & \text{if } V_{Ex} < 0 \text{ and } V_{Ey} = 0, \\ 3.1415 & \text{if } V_{Ex} > 0 \text{ and } V_{Ey} = 0, \\ 1.5707 & \text{if } V_{Ex} = 0 \text{ and } V_{Ey} > 0, \\ 4.7123 & \text{if } V_{Ex} = 0 \text{ and } V_{Ey} < 0, \\ -\arctan \frac{y_{dist}}{x_{dist}} & \text{if } V_{Ex} < 0 \text{ and } V_{Ey} > 0, \\ \arctan \frac{y_{dist}}{x_{dist}} + 1.5707 & \text{if } V_{Ex} > 0 \text{ and } V_{Ey} > 0, \\ \arctan \frac{y_{dist}}{x_{dist}} + 3.1415 & \text{if } V_{Ex} > 0 \text{ and } V_{Ey} < 0, \\ \arctan \frac{y_{dist}}{x_{dist}} + 4.7123 & \text{if } V_{Ex} < 0 \text{ and } V_{Ey} < 0. \end{cases} \tag{4.5}$$

4.4 Steady state response analysis

The first stage in the real-time experiment is calibrating the behavior of the main devices as this defines how I/O variables change (Shen et al. 2007; Wood et al. 2005; Rembe & Muller 2002). The calibration of the various equipment under study in the experiment is as follows:

4.4.1 Optoelectronic position detector

The OPD (PTQ100) is calibrated for the displacement distance range of −2 mm to +2 mm in terms of error signal −10 V to +10 V. The calibration curve of the OPD for x- and y-channel displacements are shown in Fig. 4.4. The important dynamic observations from the calibration result are:

1. Since maximum light falls on OPD, V_{Ref} = 10 V
2. OPD almost maintains linearity throughout the beam displacement regime
3. The actual results fit to 99.6 % with the expected results

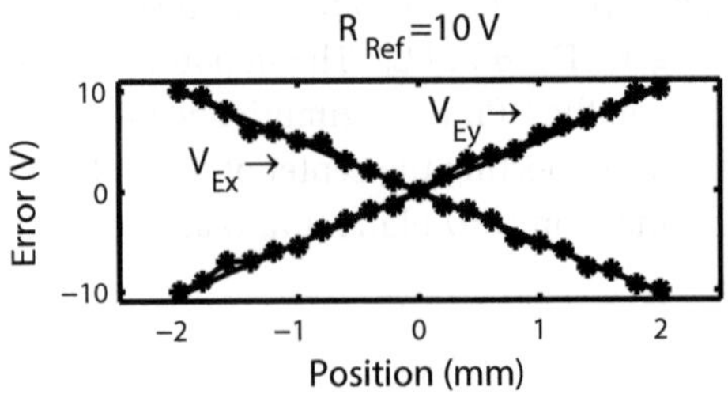

Fig. 4.4: Sensitivity measurement and calibration of OPD-PTQ100 for the beam position drift from center left to center right on the x-axis (V_{Ex}) and from center top to center bottom on the y-axis (V_{Ey}) with curve-fit model.

4. The OPD response for the x- and y-channel displacement are unique (slight deviation is seen)
5. The calibration results form a regression equation (gradient magnitude = 5.1) with the x- and y-channel intercept of −0.12 and 0.17 respectively
6. Therefore, the error of 0 V is not observed at the (0, 0) mm and
7. An error value of ±10 V is measured at the displacement distance along the x- and y-axes of ±2 mm.

4.4.2 Piezo driving amplifier

The piezo driver module (E-616) is ideally suited to amplify the arbitrary control signals (C_x and C_y) which vary in the range of −7 V to +7 V with the z-channel (z_{offset}) variation of −3 V to +3 V in this experiment and operated in open loop mode. The driver module transforms the input variations to the −20 V to +120 V on both channels with the fixed offset values ($z_{offset} = 0.5492$ V). The calibration results of the driver module are shown in Fig. 4.5 (a) and (b) and the important observations are:
1. The driver behavior is nonlinear
2. The actual results approximately fit with the expected results
3. The calibration results (best fit) forms a quadratic equation
4. Behavior for x- and y-channels are slightly unique and
5. Feedback is important to settle the control variables at the setpoint.

4.4.3 Piezoelectric actuators

Piezoelectric actuators become important in today's various applications and have many advantages such as:
1. No moving parts
2. Producing large output force
3. Fine steering resolution
4. High efficiency
5. High bandwidth and
6. Fast response time, thus gaining popularity in many micro- and nano-positioning applications.

However, the piezoelectric materials exhibit inherent nonlinearities such as creep and hysteresis (Shen et al. 2007; Wood et al. 2005; Xie et al. 2009). The S-325-Piezo-Z/Tip-Tilt platform offers piston movement of up to 30 μm (for path length adjustment) and optical beam deflection up to 10 mrad at a resolution of 50 nrad. The zero friction piezo drives and flexure guidance allow subnanometer resolution and submicro radian angular resolution. The piezo tip/tilt platform is calibrated with a 100 mm optical path

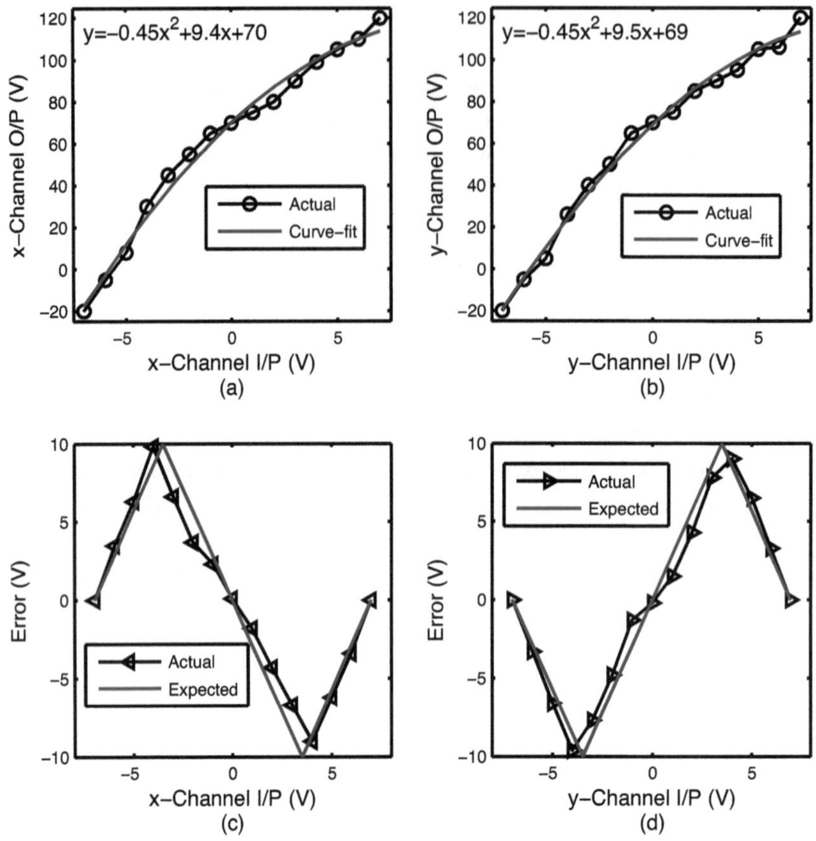

Fig. 4.5: Open loop sensitivity measurement and calibration of piezo amplifier for (a) x-channel (b) y-channel with curve-fit models and piezo platform for (c) x-channel from left-hand side to right-hand side and (d) y-channel from top to bottom.

on a vibration damped optical breadboard to undergo 100 % deflection on both axes for the control input of −20 V to +120 V. The manipulated variables vary from −7 V to +7 V while maintaining the z_{offset} constant and the results are shown in Fig. 4.5 (c) and (d). The main observations from the calibration results are:

1. The correlation between the expected and actual (measured) values is less
2. The relation between control variables and steering angle are nonlinear
3. The error variable values of (0, 0) V are not observed for the control variable values of (0, 0) V
4. Asymmetric mirror responses are seen on the x- and y-channels for the control variables step change and
5. The optimal control variable values for error voltage of (0, 0) V and receiving the maximum intensity i.e. $V^*_{Rec} = 0.3162$ V at the photodiode are $C^*_x = 0.2157$ V and $C^*_y = -0.1634$ V while z_{offset} $C^*_z = 0.5492$ V.

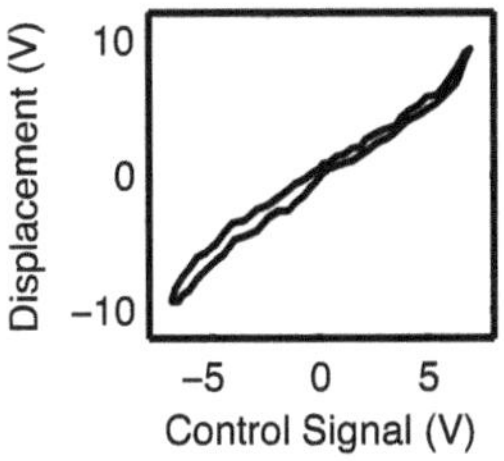

Fig. 4.6: One pattern of measured hysteresis loop obtained for cyclic variation of control signal in process nonlinearity testing.

The hysteresis characteristics and step response of piezoelectric actuators are generally nonlinear and usually unknown (Shen et al. 2007; Wood et al. 2005; Xie et al. 2009). This nonlinear behavior limits system performance via undesirable oscillations and instability. Therefore, it is difficult to operate at an accurate trajectory using conventional controllers and a feed-forward neural network is put forward to resolve hysteresis nonlinearity (Xie et al. 2009). The nonlinear hysteresis behavior of piezoelectric actuators is tested for cyclic increase and decrease of the control signal and the hysteresis loop is shown in Fig. 4.6.

As the graph shows, the steering goes to the saturation, inverse and dislocated effects. Therefore, the control variables occur together in a beam steering process on both axes and combined efforts have to be made to process each of the process variables simultaneously to meet the performance objective of the steering process. The complexity in control of beam steering arises due to their geometrical configuration, the physical phenomena present in the piezo-actuator operations and the large number of other variables involved in its operation. The nonlinear static output mapping of the system is not one-to-one (Xie et al. 2009). Further, the intelligent controller yields better nonlinear input-output mapping through which the problem is addressed in this chapter.

4.5 Development of response surface models

One of the most common approaches to obtain better results from the controller is data normalization. Data normalization can also speed up processing time for each feature within the same scale and is especially important for function approximation, modeling and control applications. The experimental input values are rescaled to lie within a range of −1 to +1. The rescaling (normalization) is accomplished by the linear interpolation formula given in Eq. (4.6). This normalization has the advantage of exactly preserving all relationships that do not introduce any bias.

$$\begin{aligned} \sigma_{Ex} &= \frac{2048 - ((V_{Ex} + 10)/0.00489)}{((V_{Ref} + 10)/0.00489) - 2048} \\ \sigma_{Ey} &= \frac{((V_{Ey} + 10)/0.00489) - 2048}{((V_{Ref} + 10)/0.00489) - 2048} \end{aligned} \tag{4.6}$$

where, σ_{Ex} and σ_{Ey} are normalized error values for x and y position displacement respectively.

The error and the corresponding control variables are experimentally measured using the developed experimental setup. To perform the experimental design, two factors at nine levels of normalized input variables (−1 : 0.25 : 1) with corresponding output values (−7 to +7 V) of two factors are used in the Design Of Experiment (DOE). An appropriate orthogonal array for the experiments must have degrees of freedom greater than or at least equal to those for the process variables. In this study, an L_{81} (9^2) orthogonal array is used.

Experimental combinations of the error variables and the corresponding control variables are the input to DOE to develop the RSM using the Taguchi modeling techniques (Asokan et al. 2008) and sample data are given in Tab. 4.1. The developed RSM equations are:

1. Linear model

$$C_x = b_0 + b_1\sigma_{Ex} + b_2\sigma_{Ey}\,,$$
$$C_y = b_3 + b_4\sigma_{Ex} + b_5\sigma_{Ey}\,;$$

2. Quadratic model

$$C_x = b_0 + b_1\sigma_{Ex} + b_2\sigma_{Ey} + b_3\sigma_{Ex}^2 + b_4\sigma_{Ey}^2\,,$$
$$C_y = b_5 + b_6\sigma_{Ex} + b_7\sigma_{Ey} + b_8\sigma_{Ex}^2 + b_9\sigma_{Ey}^2\,;$$

3. Interaction model

$$C_x = b_0 + b_1\sigma_{Ex} + b_2\sigma_{Ey} + b_3\sigma_{Ex\,Ey}\,,$$
$$C_y = b_4 + b_5\sigma_{Ex} + b_6\sigma_{Ey} + b_7\sigma_{Ex\,Ey}\,;\text{ and}$$

4. Full model

$$C_x = b_0 + b_1\sigma_{Ex} + b_2\sigma_{Ey} + b_3\sigma_{Ex}^2 + b_4\sigma_{Ey}^2 + b_5\sigma_{Ex}\sigma_{Ey}\,,$$
$$C_y = b_6 + b_7\sigma_{Ex} + b_8\sigma_{Ey} + b_9\sigma_{Ex}^2 + b_{10}\sigma_{Ey}^2 + b_{11}\sigma_{Ex}\sigma_{Ey}\,.$$

Tab. 4.1: Experimental design and their responses (observed values).

Trial number	Normalized error variables combination		Control variables (V)	
	x-position σ_{Ex}	y-position σ_{Ey}	x-channel C_x	y-channel C_y
1	1	1	−7	7
2	1	2	−6.9	5.4
3	1	3	−6.7	3.6
4	1	4	−6.8	1.6
⋮	⋮	⋮	⋮	⋮
80	9	8	6.6	−5.2
81	9	9	6.6	−6.7

The full model (second order equation) given in Eq. (4.7) is finally considered.

$$\begin{aligned} C_x &= \{0.0960622 - 6.79259\sigma_{Ex} + 0.0596296\sigma_{Ey} + 0.0045291\sigma_{Ex}^2 \\ &\quad + 0.197022\sigma_{Ex}\sigma_{Ey} + 0.0132564\sigma_{Ey}^2\} \\ C_y &= \{-0.078861 + 0.0194074\sigma_{Ex} - 7.02274\sigma_{Ey} + 0.0470611\sigma_{Ex}^2 \\ &\quad - 0.0298667\sigma_{Ex}\sigma_{Ey} + 0.136262\sigma_{Ey}^2\} \end{aligned} \tag{4.7}$$

4.6 Development of the neural network model

The normalized values of error variables are applied to the Input Layer (IL). The input layer nodes are connected through a weighted network to the array of Hidden Layer 1 (HL1). The weights 'w_{ji}^1' are linear transmission factors and bias 'b_j^1' determined from the training process and they comprise the HL1 of the proposed neural-controller. Similarly, weights 'w_{kj}^2' and bias 'b_k^2' comprise the HL2. The outputs of the HL2 are passed through another set of weighting functions 'w_{ik}^3' with bias 'b_i^3' to the Output Layer (OL). The number of hidden layers and neurons are determined through a trial and error method in order to accommodate the converged error within the goal. The 2–12–9–2 (2 neurons in the IL, 12 neurons in HL1, 9 neurons in HL2 and 2 neurons in the OL) is the structure of the developed neural-controller as shown in Fig. 4.7. The Mean Square Error (MSE) as given in Eq. (4.8) is used to measure the performance metric of the neural-controller in the training phase (Hagan & Demuth 1999; Xie et al. 2009).

$$\text{MSE} = \frac{1}{2}\sum_{j=1}^{N}(T_{pj} - O_{pj})^2\,, \tag{4.8}$$

where T_{pj} is target (desired) value from j^{th} neuron, O_{pj} is actual output value of jth neuron, and N is number of output neurons. In back-propagation, the gradient descent algorithm in which the neural network weights and biases are moved along the negative gradient of the performance function (Song et al. 2008). This rule is based on the simple idea of continuously modifying the strengths of the input connections to reduce the difference between the desired and actual output of the neural network (Hagan & Demuth 1999). This rule changes the synaptic weights in such a way as to minimize the MSE of the neural network (Ahmed & Kazuhika 2013). This rule is also referred to as the Least Mean Square (LMS) learning rule (Hagan & Demuth 1999). The fundamental mathematical equations related to the feed forward-multilayer perceptron design can be found in Hagan and Demuth (1999) and Xie et al. (2009).

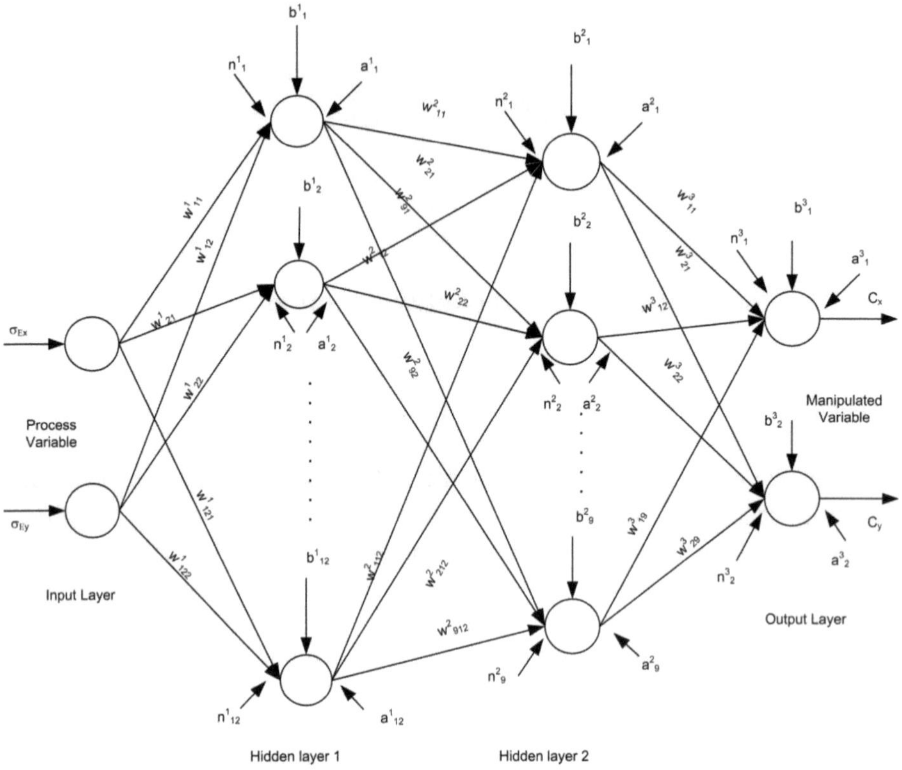

Fig. 4.7: Proposed structure of neural-controller with 2–12–9–2 multilayer perceptron model.

The back-propagation learning algorithm for the neural network structure shown in Fig. 4.7 is reviewed below:

1. Create neurons and derivative functions $f(x)$ and $f'(x)$.
2. Assign values for learning rate, momentum, goal, etc.
3. Assign smaller random value for weights and biases:

$$
\begin{gathered}
w^1_{11} \dots w^1_{121}, \quad w^1_{12} \dots w^1_{122}, \\
w^2_{11} \dots w^2_{112}, \quad w^2_{21} \dots w^2_{212}, \quad w^2_{31} \dots w^2_{312}, \\
w^2_{41} \dots w^2_{412}, \quad w^2_{51} \dots w^2_{512}, \quad w^2_{61} \dots w^2_{612}, \\
w^2_{71} \dots w^2_{712}, \quad w^2_{81} \dots w^2_{812}, \quad w^2_{91} \dots w^2_{912}, \\
w^3_{11} \dots w^3_{19}, \quad w^3_{21} \dots w^3_{29}, \\
b^1_1 \dots b^1_{12}, \quad b^2_1 \dots b^2_9, \quad b^3_1 \text{ and } b^3_2
\end{gathered}
$$

4. While con = 1, pass the training data to compute n^1_j as

$$
n^1_j = \sum_{ji} w^1_{ji} x_i + b^1_j, \quad j \to 1 \text{ to } 12, \; i \to 1 \text{ to } 2
$$

5. Apply n_j^1 to the activation function of HL1 for

$$a_j^1 = f(n_j^1)$$

6. Pass the a_j^1 to compute n_k^2 as

$$n_k^2 = \sum_{kj} a_j^1 w_{kj}^2 + b_k^2\,, \quad k \to 1 \text{ to } 9$$

7. Apply n_k^2 to the activation function of HL2 for

$$a_k^2 = f(n_k^2)$$

8. Pass the a_k^2 to compute the n_i^3 as

$$n_i^3 = \sum_{ik} a_k^2 w_{ik}^3 + b_i^3$$

9. Apply n_i^3 to activation function of OL to get a_i^3 as

$$a_1^3 = f(n_1^3) \quad \text{and} \quad a_2^3 = f(n_2^3)$$

10. Compare a_1^3 and a_2^3 with target to compute the error as

$$\Delta_i^3 = (t_i - a_i^3) f'(n_i^3)$$

11. Back-propagate the Δ_i^3 to HL2 and compute δ_k^2 as

$$\delta_k^2 = \sum_{ki} \Delta_i^3 w_{ik}^3$$

12. Calculate the error information at HL2

$$\Delta_k^2 = \delta_k^2 f'(n_k^2)$$

13. Back-propagate the Δ_k^2 to HL1 and compute δ_j^1 as

$$\delta_j^1 = \sum_{jk} \Delta_k^2 w_{jk}^2$$

14. Calculate the error information at HL1

$$\Delta_j^1 = \delta_j^1 f'(n_j^1)$$

15. Compute weight and bias correction terms as

$$\Delta w_{ji}^1 = \alpha \Delta_j^1 x_i\,, \quad \Delta w_{kj}^2 = \alpha \Delta_k^2 a_j^1\,, \quad \Delta w_{ik}^3 = \alpha \Delta_l^3 a_k^2\,,$$
$$\Delta b_j^1 = \alpha \Delta_j^1\,, \quad \Delta b_k^2 = \alpha \Delta_k^2\,, \quad \Delta b_{il}^3 = \alpha \Delta_j^1$$

16. Update the weights and biases by

$$\begin{gathered} w^1_{ji}(n+1) = w^1_{ji}(n) + \Delta w^1_{ji}\,, \quad w^2_{kj}(n+1) = w^2_{kj}(n) + \Delta w^2_{kj}\,, \\ w^3_{lk}(n+1) = w^3_{lk}(n) + \Delta w^3_{lk}\,, \\ b^1_j(n+1) = b^1_j(n) + \Delta b^1_j\,, \quad b^2_k(n+1) = b^2_k(n) + \Delta b^2_k\,, \\ b^3_i(n+1) = b^3_i(n) + \Delta b^3_i \end{gathered} \tag{4.9}$$

17.
```
if error <= goal, con = 0; disp( ); plot( ); break;
else epoch = epoch + 1; repeat training;
end,
end.
```

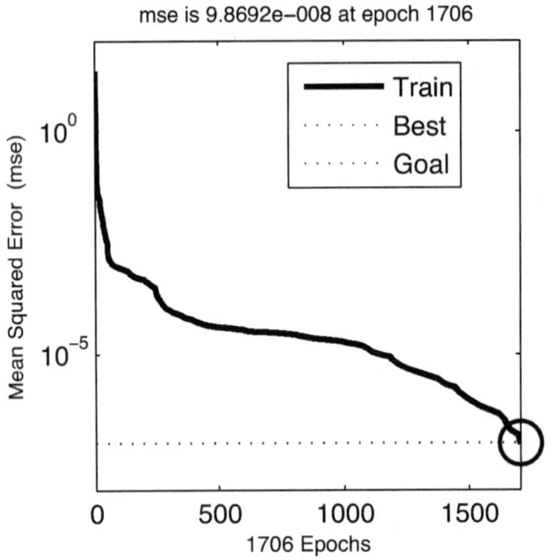

Fig. 4.8: Error rate for the learning function (error rate versus iteration). The learning procedure is stopped when the final error is below the goal.

The activation functions are maintained as identical which leads to the conclusion that the only parameters for mapping 'T_{pj}' to 'O_{pj}' are weights and biases as given in Eq. (4.9). Specifically, three weight matrices and three bias matrices are to be adjusted so that the error vector value is minimized. Thus, layered neural networks look like versatile nonlinear mapping systems with weights and bias serving as parameters. With a learning rate of 0.57 and a momentum term of 0.9, the network is trained for 1364 iterations. The MSE is less than 0.0000009427 at the end of the training process. The goal of 0.000001 is the stopping criterion for this network. The error minimization during the learning procedure is shown in Fig. 4.8. The proposed neural-controller is trained and tested in the MATLAB environment for several data patterns. One of the trained patterns of the neural-controller network structure is shown in Fig. 4.9.

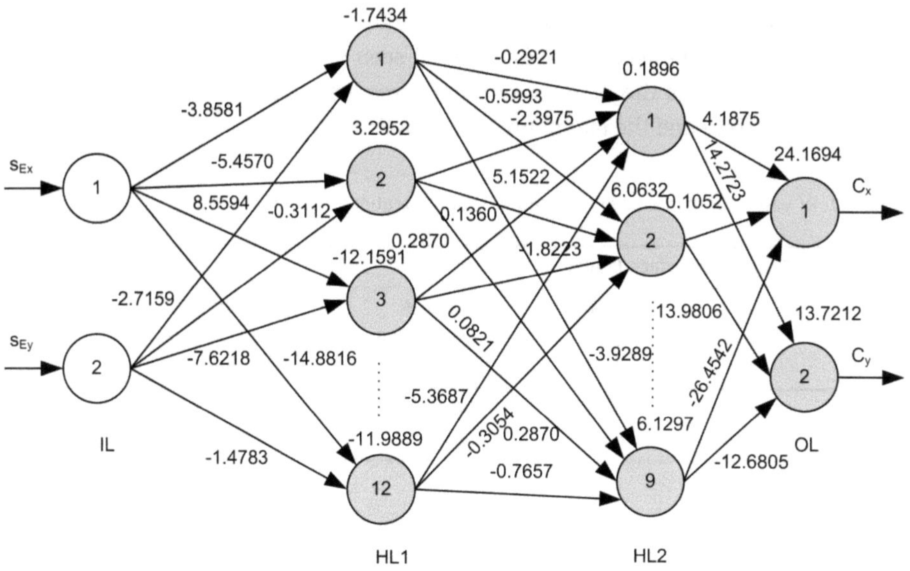

Fig. 4.9: One pattern of neural network structure with coded weights and bias values obtained from back-propagation training algorithm.

4.7 Experimental results and discussion

The RSM and neural-controller are used at the FSO link receiver setup as an observer for beam stabilization and the performances are investigated intensively. Some of the important data experimentally acquired in real time are presented and on this basis the potential and feasibility of the proposed full model and neural-controller are discussed. The outstanding performance of the neural-controller in mitigating the beam wandering and reducing the receiver signal fluctuation are analyzed. The significance of the proposed controlling scheme in improving the reliability of an FSO link is evaluated and the results are presented in this section.

4.7.1 Verification and validation of RSM and neural-controller model

In the initial condition, the control signals are biased with respect to (0.2157, 0.1634) V and necessarily the beam position must be (0, 0) mm for the controller to work satisfactorily (Perez-Arancibia et al. 2012). The experimental trials are conducted in real time over a longer period to validate the performance of the proposed control schemes (full model and neural-controller) to the position error changes. The coefficient of determination (R^2) values obtained from the full model's ANOVA table and neural-controller for the x- and y-channel control signal are given in Tab. 4.2. It can be clearly seen that, in RSMs, the full model gives better approximations for the control response surface considered in this work. The full model as given in Eq. (4.7) is used. However, the maximum values of R^2 seen using the full model are 96.63 % and 97.82 %. Whereas the neural-controller gives very accurate approximations for the control surface and the maximum values of R^2 are 99.79 % and 99.86 %.

Tab. 4.2: R^2 values of response surface models and neural-controller for C_x and C_y.

Control channel	Response Surface Model				Neural-controller
	Linear	Linear + square	Linear + interaction	Full quadratic (linear + square + interaction)	
C_x	92.09 %	93.31 %	94.41 %	96.63 %	99.79 %
C_y	91.67 %	94.21 %	95.09 %	97.82 %	99.86 %

Tab. 4.3: Results from confirmatory experiment (V_{Ref} = 10, 5, 8 V).

Trial No.	Process variables (V)		Predicted values (V)			
			Response Surface Model		Neural-controller	
	x-position V_{Ex}	y-position V_{Ey}	x-channel C_x	y-channel C_y	x-channel C_x	y-channel C_y
1	4.2678	−2.6814	2.0724	2.1267	2.4324	2.3134
2	−7.1643	3.5186	−5.9842	−2.9391	−6.4321	−2.8931
3	3.5781	−4.2716	4.0172	6.9313	4.3215	6.4512

The considered full model and neural-controller control performance are validated from confirmatory tests. In confirmatory tests, the new sets of inputs are applied and the responses are studied. Table 4.3 shows the outcome of confirmatory tests. The normal probability plot generated by the general regression graph of the residual for the control signals using the full model are shown in Fig. 4.10. The neural-controller normal probability plot of the residual generated by the MATLAB for the control signals is shown in Fig. 4.11. The scattered residuals fall on a straight line which implies that the errors are distributed normally and they have no obvious pattern and no defined structure. The residual scattering varies from the $-4.32 \cdot 10^{-3}$ to $4.01 \cdot 10^{-3}$ which is very small and the predicted values to the error changes are sufficiently equal to the observed values.

The effectiveness of the disturbance (error) rejection response for multiple setpoint errors is tested in terms of output control signal in open loop control configurations as in Fig. 4.12 (a) and (b). The expected control signals, also called reference signals, for multiple errors setpoints on the x-channel are 3.8, −2.3, 1.6 V at 0.6, 1.4, 1.8 ms respectively and on the y-channel are 2, 5.3, −1.6 V at 0.6, 1.4, 1.8 ms

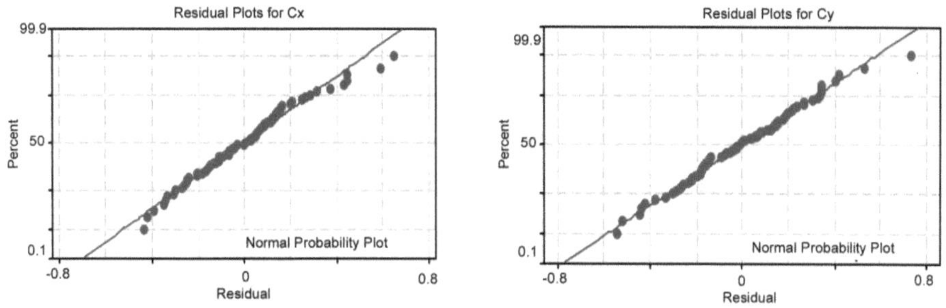

Fig. 4.10: Normal probability plot of considered full model: percent versus residual for x- (left) and y-channel (right) control.

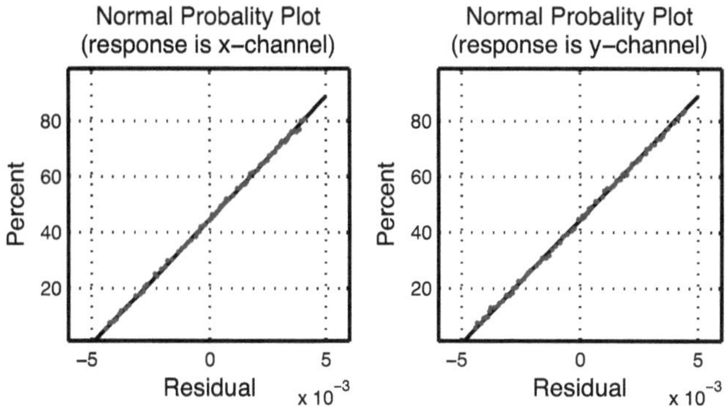

Fig. 4.11: Normal probability plot of proposed neural-controller: percent versus residual for x- (left) and y-channel (right) control.

respectively. Since the accurate state of the physical system is unavailable for measurement (Fidler et al. 2010), the disturbance inputs are given at the transmitter, and the corresponding control outputs are measured at the receiver and compared with the reference signal. The timing behavior of the proposed control schemes in terms of accuracy and speed with respect to the reference signal is shown in Fig. 4.12 and evidence that the neural-controller rejected the disturbance more effectively than the RSM controller.

The neural-controller response rising and deadtime is very small, processing gain is very high and does not lead to overshoot, undershoot or oscillatory effects, whereas the RSM controller exhibits all these effects before settling at zero error (setpoint). It is observed with these results that the neural-controller actions against the error changes are accurate and more rapid. The percentage of error predicted by the developed models is calculated by

$$\frac{\text{Observed value} - \text{Predicted value}}{\text{Predicted value}} \cdot 100$$

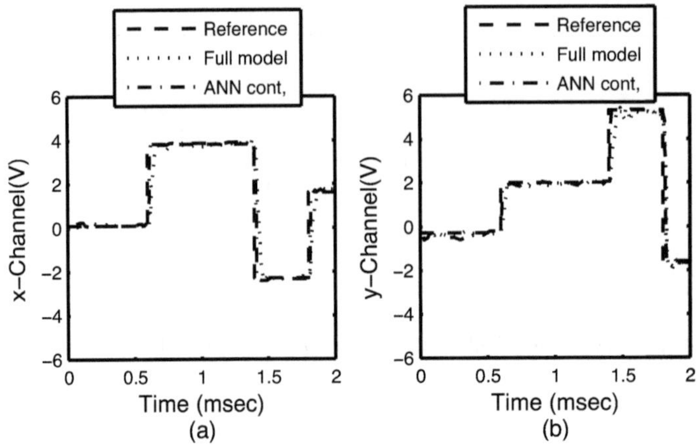

Fig. 4.12: Open loop plant response for multiple setpoint errors: Reference control voltages are (a) 0.2157 → 3.8 → −2.3 → 1.6 V for the x-channel and (b) −0.1634 → 2 → 5.3 → −1.6 V for the y-channel.

as given in Tab. 4.4. The maximum percentage of prediction error observed from the open loop tests are shown in Fig. 4.13.

Tab. 4.4: Validation test results of developed response surface model (full) and neural-controller for C_x and C_y.

Trial No.	Response surface model				Neural-controller			
	Value (V)		Error in %		Value (V)		Error in %	
	C_x	C_y	C_x	C_y	C_x	C_y	C_x	C_y
1	1.3756	3.1156	5.5903	0.3819	1.4524	3.1276	0.0069	−0.0032
2	−2.9997	−5.0001	14.2714	3.5619	−3.4275	−5.1778	0.0088	0.0077
3	6.2845	−4.5006	−0.2021	7.0124	6.2711	−4.8160	0.0112	0.0042

The percentage of prediction errors obtained from the neural-controller throughout the test space is almost very close to 0 % and the min/max values are −0.1213 % and 0.1342 %.

This control performance clearly shows that the neural-controller has a good capability to track and stabilize the beam to a particular point, further good beam stabilization is obtained even for the disturbances introduced by the z_{offset} changes. The neural-controller takes almost close corrective action so as to bring back the errors to (0, 0) V. Thus the neural-controller predicts the exact control surface and hence it is only used at the receiver to control the beam wandering throughout the experimentations.

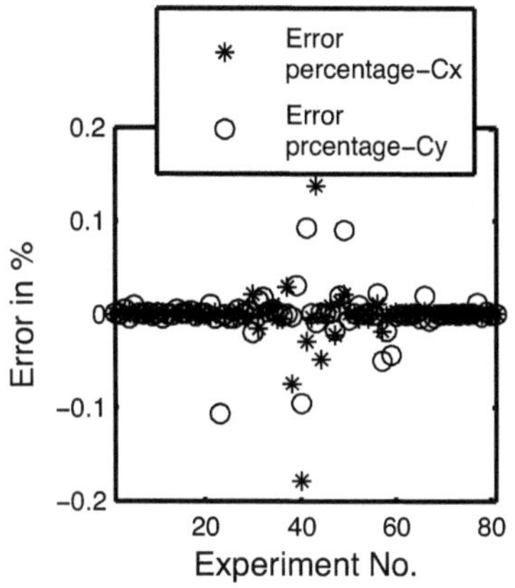

Fig. 4.13: Percentage of error of control signal predicted by the neural-controller in open loop configuration for x- and y-channels.

4.7.2 Behavioral study of neural-controller in beam alignment

The performance improvement in beam stabilization through the beam steering control system is tested in closed loop in real time with controller turned-off and turned-on conditions. The beam centroid coordination on the 2D plane of OPD is estimated using Eqs. (4.3)–(4.5) and the compass plot for a portion of real-time experimental data is shown in Fig. 4.14 (left).

The radial distance is calculated using Eq. (4.4) in no control and in control conditions and their corresponding histogram plot is also shown in Fig. 4.14 (right) which evidences that the beam steering control system mitigates beam wandering and stabilizes the beam almost at the center of the photodiode. The sample count occupies all the levels of radial distance in no control, whereas in control, the maximum count value is around 0 mm. The interesting observations from Fig. 4.14 in no control are:

1. Beam centroid wanders in almost all the directions on the detector plane
2. Maximum displacement magnitude of 2 mm is seen
3. Centroid movements are highly random
4. Sometimes the centroid goes out-off (outage) the detector plane i.e. the FSO link is disconnected
5. A wide range of power fluctuation at the photodiode is seen.

In control are:

1. Centroid movements are controlled and are always around the optimal point
2. Maximum displacement magnitude of 0.42 mm is seen for a short period
3. Very narrow range of power fluctuations are seen
4. Maximum intensity of beam (Gaussian beam) is concentrated at the center of the photodiode for a long period.

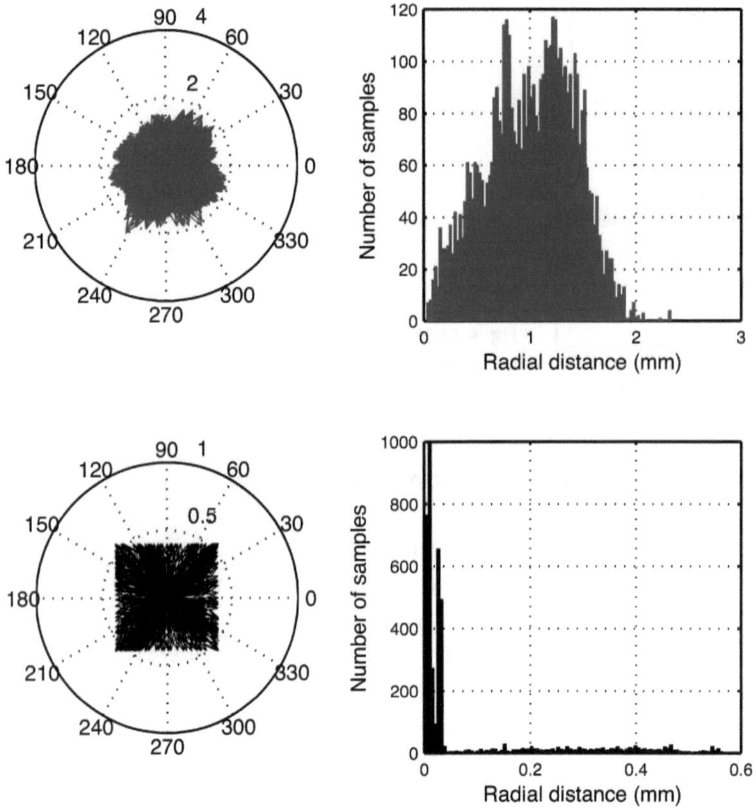

Fig. 4.14: Figures on the left show compass plot of time series laser beam spot centroid motion on OPD with beam steering control off and on conditions. Figures on the right show the corresponding histogram of radial distance (from plane center to beam centroid).

4.7.3 Analysis of receiver signal quality improvement

The (0, 0) V outputs from the OPD mean that the beam centroid is positioned on the center of the photodiode and its output is exactly the minimum possible value i.e. should coincide with the lower bound of the photodiode output $V^*_{Rec} = 0.3162$ V. In the absence of sensor noise, the photodiode output difference ($V_{Rec}(t) \sim V^*_{Rec}$) measured in volts is assumed as directly related to beam wandering. This effect is mitigated using the beam steering system, so that beam stability on the photodiode is improved. The power fluctuations of the received signal are evaluated with and without the beam steering system. As a direct consequence of the disturbance rejection in each channel, the average value of the photodiode output is significantly closer to the optimal value V^*_{Rec}, despite the fact that $\{\sigma_{Ex}(t), \sigma_{Ey}(t)$ and $C_x(t), C_y(t)\}$, $\forall t$. It can be clearly seen from Fig. 4.15 (left) that the received signal fluctuates with the broader variations from 0.3162 V to 1.2361 V in the outer scale when the beam steering is turned off and

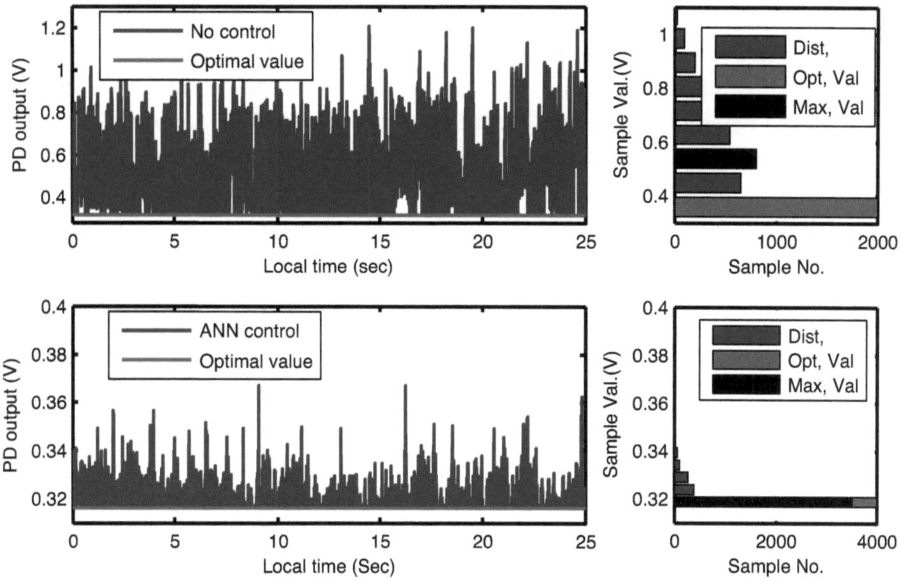

Fig. 4.15: Figures on the left show the time series of photodiode output without (top) and with (bottom) beam steering and on the right show the corresponding histogram.

within the inner scale with the min/max values of 0.3162 V and 0.3716 V when the beam steering is turned on. The histogram plot shown in Fig. 4.15 (right) illustrates that the photodiode output fluctuations are randomly distributed at almost all the levels due to the beam wandering in no control and in control it is arrested and the maximum count value is seen for the optimal value for a longer period.

These results exhibit the efficiency of the neural-controller in steering the beam as required. Therefore the neural-controller in a closed loop steering system improves the reliability of an FSO link. The receiver signal fluctuation, even with the beam steering turned-on condition, reaches the maximum value of 0.3716 V. This is due to purely optical power fluctuation in the transmission path (refractive index of channel, Mie/Rayleigh scattering, absorption and particle interactions with the optical beam i.e. path loss) which could not be controlled by the steering system and is a matter for separate research for respective local meteorological changes.

4.8 Summary

This chapter presents a novel approach for beam steering in FSO links. The significance of beam steering in FSO links is discussed and the related literature review results are presented. The construction of a terrestrial FSO link for the range of 0.5 km at an altitude of 15.25 m with a system dynamics study and calibration of the optoelectronic equipment are explained. The response surface model of linear, quadratic, interaction and full (linear + quadratic + interaction) are developed using Taguchi's method (DOE) and the maximum coefficient of determination of 96.63 % and 97.82 % are achieved using the full model for the response signals. The approach of neural-control design in the MATLAB environment is presented and the optimal weights and biases (goal $\leq 10^{-6}$) are obtained with a 2–12–9–2 structure neural network of multi-layer perceptron model. The information related to the beam displacement and signal power at the photodiode are transferred (logged) to the computer and on/off line statistical analyses are carried out. The validation and confirmatory test for full model and neural-controller are conducted and the achieved min/max values of magnitude of percentage of prediction error are −0.2021 % and 17.5827 %, and −0.1213 %, and 0.1342 % respectively. A best control response is achieved from the neural-controller rather than the full model controller for dynamic disturbances in closed loop conditions and the results are reported. It is suggested that the neural-controller can be used for fine steering applications instead of the complicated computational algorithms found in literature. The capabilities of mitigating beam wandering are evaluated with and without beam steering techniques. The values of the receiver signal are brought to a controlled range of 0.3162 V to 0.3716 V. The results presented in this chapter are intended for setting up an FSO communication data link transceiver. This proposed neural-control model can also be used in various applications like laser cutting, laser metrology, laser surgery, focusing, spotting, beacon positioning, beam coupling, etc. (Bazil Raj et al. 2014b, with permission of OSA).

5 Low power and compact RSM and neural-controller design for beam wandering mitigation with a horizontal-path propagating Gaussian-beam wave: focused beam case

Beam wander on the detector plane is one of the main causes of major power loss which severely degrades the performance of Free Space Optical (FSO) links. Confronted with this big problem, designing a suitable controller to compensate beam wandering at a fast rate so as to increase beam stability becomes significant. This chapter presents an investigation of the performance of two types of controller designed for increasing the stability of the beam on the detector plane under dynamic disturbances. The first design is based on Taguchi's method: Response Surface Model (RSM) controller while the second is the Artificial Neural Network (ANN) method (neural-controller). These controllers process the beam spot information and generate the necessary outputs to mitigate beam wandering, so as to perfectly couple the Power In the Bucket (PIB): receiver aperture, into the detector. Pipelined-parallel architecture for both controllers are proposed and developed in a Field Programmable Gate Array (FPGA). The implementation of these two candidate controllers is described in detail as installed at the receiver station. Evidence of the suitability and the effectiveness of the proposed controllers in terms of prediction exactness, prediction error, reduction of beam wander, response to impulse and effective scintillation index are provided through experimental results from the FSO link established for the horizontal range of 0.5 km at an altitude of 15.25 m.

5.1 Introduction

In recent years, Free Space Optical (FSO) links have received much attention from both industry and academia. Especially when the Radio Frequency (RF) spectrum is heavily congested, the features of a license-free service become a significant advantage of an FSO link (Vu et al. 2013). In an FSO link the modulated beam travels through a channel which is either totally the atmosphere, such as in the case of terrestrial and airborne communication, or at least partly through the atmosphere, such as in the case of earth to satellite communication. The fundamental limitation of an FSO link arises from the environment through which it propagates. Although relatively unaffected by rain and snow (Colvero et al. 2007), FSO link systems can be severely affected by fog and atmospheric turbulence (Sergeyev & Roggemann 2013). A problem that has to be dealt is the influence of dynamic atmospheric changes, thermal expansion, weak earthquakes, typical high-rise building sway (Tianping et al. 2005; Wright & Easom 2003) and mechanical vibration of the optoelectronics equipment like transmitting optics, telescope etc. in reference to the center of the detector, which results in variations of the Angle of Arrival (AoA) of the laser beam (Sergeyev & Roggemann 2013; Liu 2009; Ciaramella et al. 2009). Building sway causes vibration on the transmitted beam moving it out-off Line of Sight (LoS) from the direction of the receiver. These vibrations decrease the average received signal, which in turn increases the Bit Error Rate (BER) (Tsiftsis et al. 2006) and severely degrades link performance (Barsimantov & Nikulin 2011; Jiang et al. 2008).

During data transmission, the angle between the LoS and the transmit beam axis must be kept to within a fraction of the transmit beamwidth which may be as small as a few microradian (Killinger 2002; Tsou & Yan 1999). Existence of a LoS and continuous alignment between the communicating optical antennas is one of the key requirements for successful installation of an FSO link system (Lee et al. 2001; Yoshida et al. 2009). Designing a suitable control system for beam aligning, tracking and positioning to significantly improve the beam stability on the receiver plane becomes significant (Sergeyev & Roggemann 2013; Yoshida et al. 2009; Perez-Arancibia et al. 2012). Therefore, the first and most important approach in FSO links is beam steering which mitigates beam wandering (Fielhauer et al. 2003; Suite et al. 2004; Gu & Zhu 2010) effects and is addressed with RSM and intelligent decision-making techniques in this chapter. The complexities of controller design for such process increases since most of the processes are inherently nonlinear thereby making the control of a nonlinear system nontrivial (Hsu 2008). The Artificial Neural Network (ANN) is the most popular intelligent tool due to its high ability to learn from sample data and to yield solutions for new data (Xie et al. 2009; Yu et al. 2005; Maeda & Tada 2003). The hardware implementation of a neural network becomes important to operate the beam steering control process at a faster rate (Valle & Diotalevi 2004; Bade & Hutchings 1994) and in a parallel manner (Gago et al. 2007; Le & Jeon 2010; Pasupathi et al. 2014). The beam steering controller is attempted using (i) Taguchi's method: Response Surface Model (RSM) and (ii) ANN: neural-controller. In this work, both controllers are implemented in a Field Programmable Gate Array (FPGA) and the control performances are evaluated.

The rest of the chapter is organized as follows: Section 5.2 presents the background and related works, Section 5.3 describes the experimental plant and displacement error computation, Section 5.4 explains the implementation of the RSM controller, Section 5.5 describes the novel digital architecture designed for implementing the neural-controller with Universal Asynchronous Receiver Transmitter (UART) architecture, Section 5.6 discusses the experimental results and data analysis, and Section 5.7 draws some conclusions.

5.2 Background and related works

In the last few years, many of controller designs related to laser beam aligning, tracking and positioning have been investigated out and can be found in the literature. The publications/results closely related to our work are discussed in this section.

Two optimization models are established and solved by Xian Liu (2009) for continuously adjusting the transmitter power, wavelength and telescope gain to mitigate the beam LoS deviations due to building sway and atmospheric interference.

Ciaramella et al. (2009) present a dedicated electronic control unit that effectively tracks signal beam wandering caused by atmospheric turbulence and mechanical vibrations in the wireless optical data link of 212 m between two buildings in Pisa, Italy. The tracking beam is coupled with the single mode fiber.

Shinhak Lee et al. (2001) present a small indoor experimental setup for accelerometer (instead of Optoelectronic Position Detector (OPD)) assisted beam (beacon laser) tracking and pointing of optical beam (transmit laser) for deep space optical communications. Setup vibrations are introduced by the piezo-actuators and the transmit laser beam is steered by the Fast Steering Mirror (FSM).

Nestor O. Perez-Arancibia et al. (2012) developed the indoor beam stabilization system with the observer-based feedback controller. The disturbance is generated by the FSM optical shaker and corrected by the FSM using the extended Kalman filter. The experimental results are presented and briefed.

Fielhauer et al. (2003) conducted experiments to compare the steering performance of macro-tip-tilt and mesoscale position beam steering transducers for FSO link applications. The position errors are measured using OPD and the results are discussed. It is shown that a robust FPGA implementation of the tracking controller will give good beam stabilization.

Suite et al. (2004) explain how to install the FSM and OPD to correct focal spot position fluctuation and resulting power loss. The experiment is carried out across the bay (20 mile round-trip) at Chesapeake Bay Detachment (CBD) of the Naval Research Laboratory (NRL). The calibration results of the OPD are given and the experimental results with and without active corrections are explained.

Wen-Fang Xie et al. (2009) propose a neural network based adaptive controller for piezoelectric actuators with unknown hysteresis. A neural network based dynamic pre-inversion compensator is designed to cancel out the effect of the hysteresis.

Gago et al. (2007) explain the tip-tilt mirror control based on the FPGA technology for adaptive optics system and report that the FPGA is the best viable platform for the control process when a huge amount of parallel processing is needed. The experiment is conducted with integer and floating point calculations. The results are presented and analyzed.

A neural-network based, low speed damping controller is designed to remove the nonlinear disturbance in an operating stepper motor by Quy Ngoc Le and Jae-Wook Jeon (2010). The back-propagation learning algorithm is used and the neural-controller is implemented in FPGA. The experimental results are investigated. Banuelos-Saucedo et al. (2003) describe the implementation of neuron models in FPGA. The VHDL codes are presented and the simulation results are compared with the real outputs.

5.3 Experimental plant configuration and centroid error computation

The FSO link transmitter and receiver setups are built on a tower (lat. 10°38′46.7334″ and long. 79°3′12.0774″) and on the existing Information Technology (IT) block (lat. 10°38′52.8468″ and long. 79°2′56.6268″) at our college respectively. The transmitter is located 0.5 km away from the IT block. The transmitter and receiver are inline at an altitude of 15.25 m. The experimental setup is established with the optoelectronic devices and mounted on vibration damped optical breadboards. AC plant, mechanical workshop, civil engineering block, playground and small crop field are the sources affecting the optical path in addition to the atmospheric disturbance effects. Therefore, the terrain type is open flat, grass, and few isolated obstacles. A Google image of the experimental location representing the transmitter and receiver is shown in Fig. 5.1. The schematic diagram of the experimental setups shown in Fig. 5.2 consists of modulatable Beta-Tx optical source, optical shaker, impulse and disturbance sequence generator circuit (in FPGA), Digital to Analog converter (D/A) and transmit optics at the transmitter and telescope, FSM, variable beam splitter, OPD, Analog to Digital converter (A/D), RSM and neural-controller (in FPGA) at the receiver. The disturbances are generated as suggested by Perez-Arancibia et al. (2012). The OPD delivers a linear response over a range limited by the beam spot size and becomes highly nonlinear

Fig. 5.1: Optical propagation path: The receiver system is located on the IT block at the left and the transmitter system is located on a tower at the right.

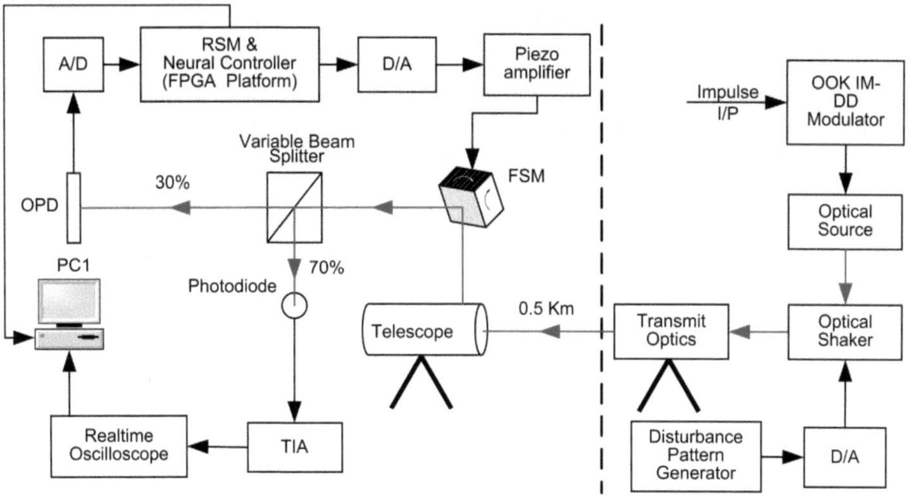

Fig. 5.2: Schematic of the free space optical communication experimental setup for 0.5 km data link. Left-hand side: receiver with beam steering optoelectronic assembly and right-hand side: transmitter (red arrow line: optical path and black arrow line: signal path).

beyond that range. Therefore, choosing the sufficient spot size as the expected angle of arrival becomes significant.

The DC input is given to the modulator circuit and the unmodulated beam is used to characterize the proposed control loop in open and closed control configurations. The unmodulated light falls on the optical shaker (secondary source of jitter) and is transmitted to the receiver through the atmospheric turbulent channel. The telescope captures all the optical power and reflects it to fall on the FSM. The incident laser beam of the FSM is reflected to the variable beam splitter that splits the incident beam into two: reflecting beam and propagating beam. The reflected beam is made to fall on the optical detector (photodiode) used for communication purposes.

The propagating beam falls on the OPD and the voltage generated by each quadrant is proportional to the optical energy illuminating its surface. The beam spot displacement errors V_{Ex} and V_{Ey} along the x- and y-channel (2D Plane) are measured using Eq. (5.1) by (Suite et al. 2004; Yu et al. 2005) as

$$\begin{aligned} V_{Ex} &= \{(V_A + V_C) - (V_B + V_D)\}, \\ V_{Ey} &= \{(V_A + V_B) - (V_C + V_D)\}. \end{aligned} \tag{5.1}$$

The reference signal V_{Ref} is measured by the algebraic sum of all signals by using Eq. (5.2) as (Suite et al. 2004, Yu et al. 2005)

$$V_{Ref} = (V_A + V_B + V_C + V_D). \tag{5.2}$$

The analog output signals V_{Ex}, V_{Ey} and V_{Ref} vary from -10 V to $+10$ V and 0 V to $+10$ V respectively. The outputs of OPD are applied to A/D (AD1674 converter core and analog multiplexer ADG509A) and then the normalized error data are given to the RSM and neural-controller developed in FPGA. The controller's output data are converted to analog signal using D/A and given to the FSM through the piezo-amplifier. The measurement data and control process performance are continuously monitored using PC1. The azimuth and elevation distance are measured by

$$x_{dist} = -2(V_{Ex}/V_{Ref}) \quad \text{and} \quad y_{dist} = +2(V_{Ey}/V_{Ref})\,. \tag{5.3}$$

The radial distance is measured by

$$\gamma = \sqrt{x_{dist}^2 + y_{dist}^2} \tag{5.4}$$

The AoA is estimated by (Yu et al. 2005)

$$\theta_x = \tan^{-1}(x_{dist}/500100) \quad \text{and} \quad \theta_y = \tan^{-1}(y_{dist}/500100)\,. \tag{5.5}$$

An impulse generator circuit is developed in the FPGA whose output is given to the modulator whenever it is enabled. Impulses are applied to the modulator with sufficient delay (off period) and the entire system impulse response is studied practically.

5.4 Formulation and implementation of direct controller

The experimental input values are rescaled to lie within a range of −1 to +1. The rescaling (normalization) is accomplished by the linear interpolation formula derived as given in Eq. (5.6). The input and response relations are experimentally measured using the experimental setup shown in Fig. 5.2. The single-precision (32 bit) floating point arithmetic standard is followed for accomplishing the normalization and arithmetic operations (Benrekia & Attari 2010; Monmasson & Cirstea 2007).

$$\begin{aligned} \sigma_{Ex} &= \frac{2048 - ((V_{Ex} + 10)/0.00489)}{((V_{Ref} + 10)/0.00489) - 2048}\,, \\ \sigma_{Ey} &= \frac{((V_{Ey} + 10)/0.00489) - 2048}{((V_{Ref} + 10)/0.00489) - 2048} \end{aligned} \tag{5.6}$$

where, σ_{Ex} and σ_{Ey} are normalized error values for the x and y position displacement errors respectively. The hardware design using Xilinx block set for the Eq. (5.6) is shown in Fig. 5.3. The individual and combined effects of input factors on the response are analyzed and subsequently direct controller mathematical models are formulated using Minitab software. Although there are many parameters that affect the steering operation in beam stabilization, x- and y-displacement errors contribute the greatest effect on the success of its operation as per the computed analytical results obtained from the analysis of variance (ANOVA): for C_x, σ_{Ex} is the dominating parameter followed by σ_{Ex}^2, σ_{Ey}, $\sigma_{Ex}\sigma_{Ey}$, and σ_{Ey}^2. The parameter σ_{Ex} contributes 53.39 %

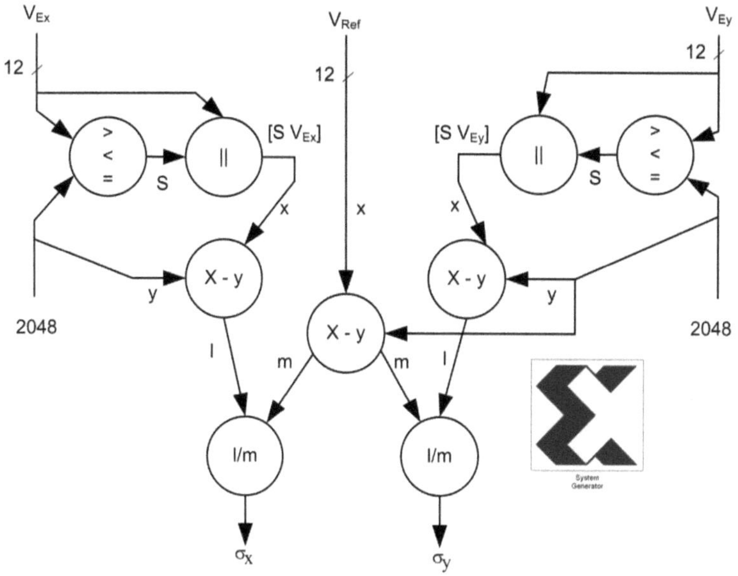

Fig. 5.3: Hardware design flow graph of data normalization algorithm in PVs preprocessor unit. Error signal V_{Ex} and V_{Ey} variations −10 V to +10 V are normalized with respect to V_{Ref}. Error signal of 0 V is equal to 800H = $(2048)_{10}$.

and σ_{Ey} contributes 0.6286 % towards the desired x-control. The effect of σ_{Ey}^2 is very small when compared to their individual and combined contributions. Similarly, the individual and combined effects of σ_{Ex} and σ_{Ey} for y-control are also analyzed: σ_{Ey} is the dominating parameter followed by σ_{Ey}^2, σ_{Ey}, $\sigma_{Ex}\sigma_{Ey}$, and σ_{Ex}^2.

It is obvious that the factor σ_{Ey} contributes 49.17 % on y-control, whereas the parameter σ_{Ex} contributes 4.215 % towards y-control. The effect of σ_{Ex}^2 is very small when compared to their individual and combined contributions. Therefore, only these parameters are considered in this work as the practical constraints along with the desired response surface. The direct controller is formulated based on the individual and combined effect of steering parameters as

$$\begin{aligned} C_x &= b_0 + b_1\sigma_{Ex} + b_2\sigma_{Ey} + b_3\sigma_{Ex}^2 + b_4\sigma_{Ey}^2 + b_5\sigma_{Ex}\sigma_{Ey}\,, \\ C_y &= b_6 + b_7\sigma_{Ex} + b_8\sigma_{Ey} + b_9\sigma_{Ex}^2 + b_{10}\sigma_{Ey}^2 + b_{11}\sigma_{Ex}\sigma_{Ey}\,, \end{aligned} \tag{5.7}$$

where

$$\begin{gathered} b_0 = 0.0960,\quad b_1 = -6.792,\quad b_2 = 0.0596,\quad b_3 = 0.0045,\quad b_4 = 0.0132, \\ b_5 = 0.1970,\quad b_6 = -0.0788,\quad b_7 = 0.0194,\quad b_8 = -7.022,\quad b_9 = 0.0470, \\ b_{10} = 0.1362,\quad \text{and}\quad b_{11} = -0.0298\,. \end{gathered}$$

Equation (5.7) is a full model equation, i.e. it consists of linear, quadratic and interaction terms of the input factors (Ahilan et al. 2013). The pipelined-parallel architecture

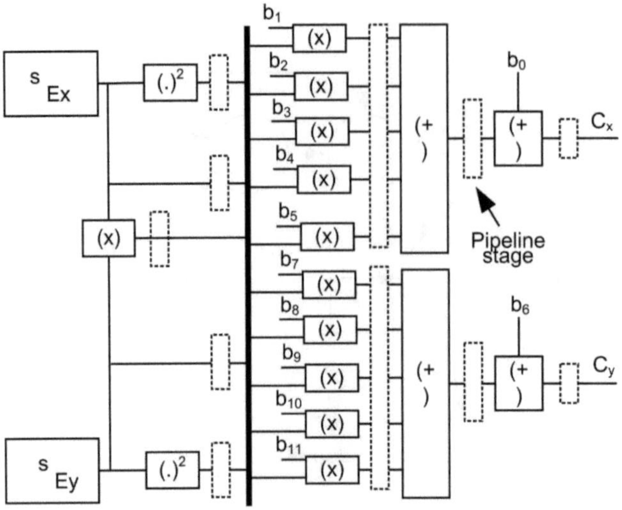

Fig. 5.4: Direct controller design based on the response surface model with pipeline registers.

designed for the direct controller implementation in FPGA is shown in Fig. 5.4. The entire operations of the controller are managed by the pipeline stages (Perreault & Bifano 2004). The direct controller outputs are forwarded to a 12 bit resolution D/A (DAC 7728 core converter) circuit.

5.5 Hardware architecture and implementation of the neural-controller

For on-chip training, hardware platforms suffer several disadvantages such as achieving high data precision, high hardware cost of calculations, development time and flexibility of the platform as compared to software (Omondi & Rajapakse 2006; Dias et al. 2004; Gomperts et al. 2011; Skoda et al. 2011). Since more computations and complicated arithmetic calculations are involved in neural-controller design, at present, most of the neural-controllers are simulated by software programs and implemented in FPGA or fabricated using VLSI technology for the applications requiring parallel and high speed computations (Omondi & Rajapakse 2006; Skoda et al. 2011). Therefore, the neural-controller weights and bias values are obtained from the MATLAB environment as a starting point for hardware implementation. The complete training and designing procedure can be found in Chapter 4. The number of hidden layers and neurons are determined through a trial and error method in order to accommodate the converged error within the goal. The 2–12–9–2 (2 neurons in the Input Layer (IL), 12 neurons in the 1st Hidden Layer (HL1), 9 neurons in the 2nd Hidden Layer (HL2) and 2 neurons in the Output Layer (OL)) is the structure of the developed neural-controller.

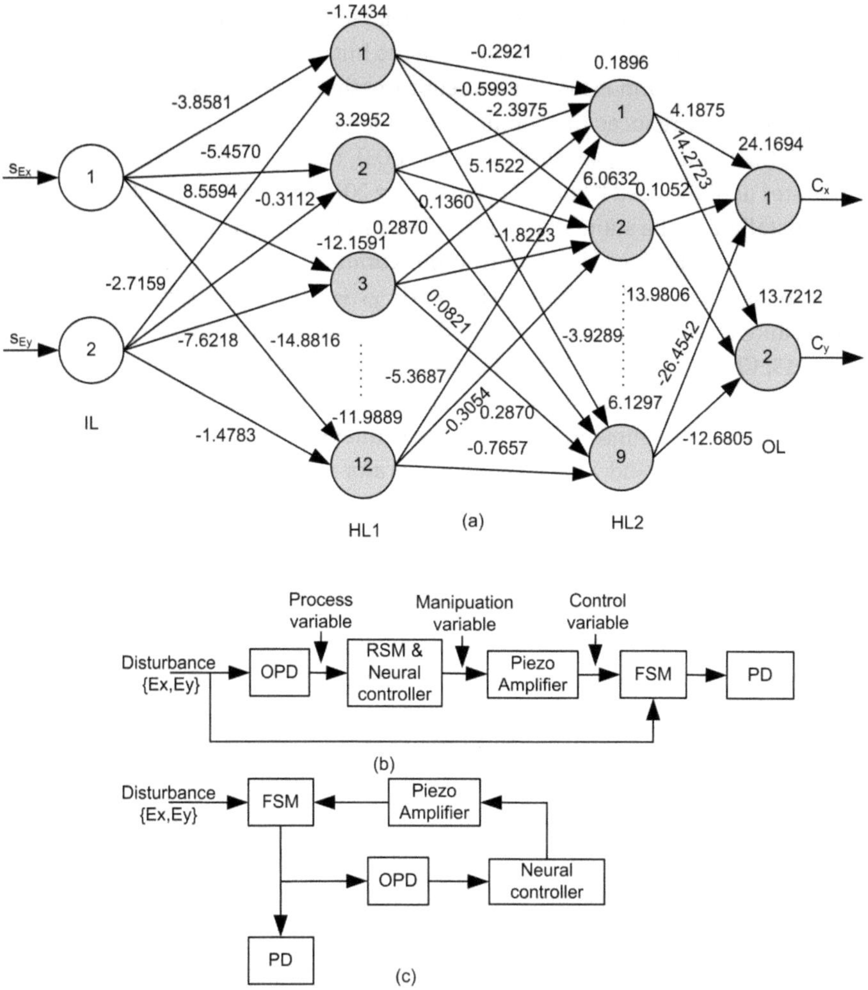

Fig. 5.5: (a) Typical back-propagation neural network trained pattern coded with weights and bias values, Standalone Linear Time Invariant (LTI) system with neural-controller in (b) open loop: incident light falls on the OPD and FSM. FSM driven signals are approximately proportional to the values of process variables, and (c) closed loop: incident light falls on FSM. The reflected light falls on the OPD and a control signal is applied in feedback control configuration.

The fundamental mathematical equations related to the feedforward multilayer perceptron design with back-propagation learning algorithm can be found in Section III of Le and Jeon (2010). With a learning rate of 0.57 and a momentum term of 0.9, the network is trained for 1364 iterations. The mean square error between desired and best is less than 0.0000009427 at the end of the training process. One of the trained patterns of the neural-controller network structure with weights and bias values is shown in Fig. 5.5 (a). A novel pipelined-parallel architecture is developed in FPGA to implement the neural-controller. The software implementation of a neural-controller usually takes a long period of time to execute the huge amount of computations. The ability of FPGA to perform massive parallel processing increases the obtainable control loop frequency, reduces the computational latency in the control system (Kepa et al. 2008; Dias et al. 2004; Meyer-Bäse et al. 2003), makes the controller stand alone and operates at high speed (Omondi & Rajapakse 2006; Meyer-Bäse et al. 2003).

The installation of a standalone Liner Time Invariant (LTI) system with a neural-controller in open and closed loop control configuration at the receiver as the observer is shown in Fig. 5.5 (b) and (c) respectively. FPGA gives great flexibility in prototyping, designing and development of complex hardware (Vasumathi & Moorthi 2012; Tisan & Cirstea 2013; Chen et al. 2012), high gate density (i.e. tens of millions logic gates) and many features necessary for neural-controller implementation (Daalen et al. 1997). The mixed design environment: Very high speed integrated circuit Hardware Description Language (VHDL) and Xilinx system generator (Tisan & Cirstea 2013; Bazil Raj et al. 2014c) are used since they simplify the development of the neural-controller and it is possible to model and simulate a digital system from a high level of abstraction module designs (Tisan & Cirstea 2013; Saunter et al. 2006). The global block diagram of the neural-controller developed based on the COordinate Rotation DIgital Computer (CORDIC) is shown in Fig. 5.6 and it consists of various main data processors and a control engine. All the units/modules are developed to operate in parallel with the controlled data and command flow. The finite state machine engine (Melay model) control information provides command signals that coordinate and execute various operations in the data processor section in order to accomplish the desired data processing task (Anbuselvi & Bazil Raj 2014).

Tab. 5.1: Example of a process-time diagram (brief) showing the pipeline of the computation process of the neural-controller with process description and time.

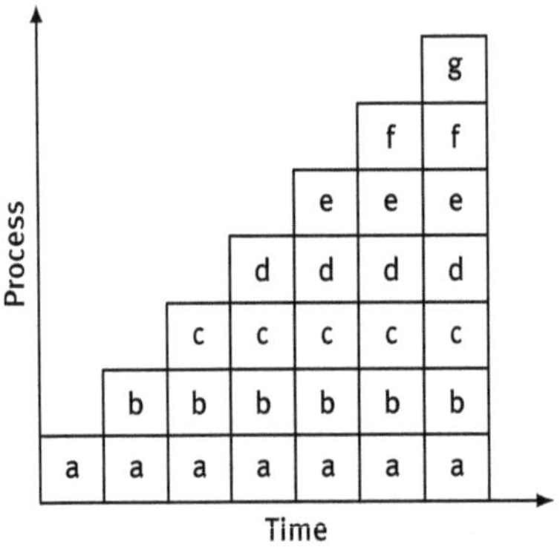

Pipeline stage	Process description	Process time
a	Process variable reading	11 u.t.
b	i/p data pre-processing	7 u.t.
c	$n^1(\cdot)$ and $a^1(\cdot)$ computation; $a^1(\cdot)$ value routing to HL2-DDRC	20 u.t.
d	HL2-MAC calculation	36 u.t.
e	$n^2(\cdot)$ and $a^2(\cdot)$ computation; $a^2(\cdot)$ value routing to OL-MAC	28 u.t.
f	OL-MAC calculation	21 u.t.
g	$n^3(\cdot)$ and $a^3(\cdot)$ computation; o/p forward control	16 u.t.

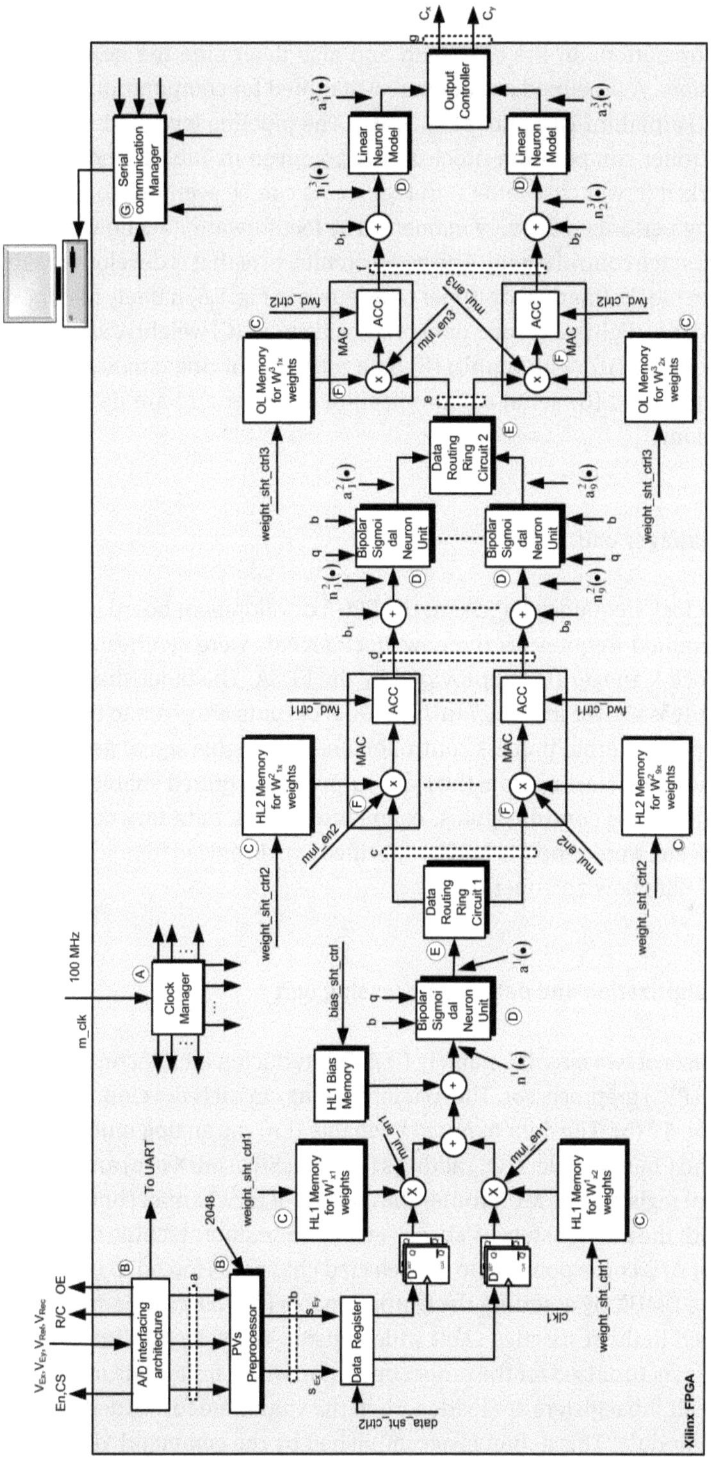

Fig. 5.6: Global block diagram of clock manager, data acquisition, neural-controller and UART implementation in the FPGA device. A, B, C, D, E, F, and G are main subcircuits; a, b, c, d, e, f, and g are pipeline stages.

The digital circuits act as the controller that provides a time sequence of signals for initiating the operations in the data path and also determine the next state of the control subsystem. A pipelined architecture is required for computational operations of algorithms (Farmahini-Farahani et al. 2010). The pipeline level and the schedules of neural-controller computation process flow are given in Tab. 5.1 and the pipeline stages are marked (lower case letters) in Fig. 5.6. As can be seen in Tab. 5.1, the operations are being performed at every moment. The feedforward data flow is controlled by the pipeline stage control signal. The novel architecture that is developed is divided into seven main blocks (marked by upper case letters in Fig. 5.6) namely (A) clock manager unit, (B) signal digitization and data processing unit, (C) weight and bias memory management circuit, (D) neuron unit, (E) data routing and ring circuit, (F) multiply-accumulator unit, and (G) serial communication manager and are described in the following sections.

5.5.1 Clock manager unit

The on-board clock frequency 'm_clk' in the FPGA development board is 100 MHz. To achieve the required frequencies the new clock signals were synthesized by means of a Digital Clock Manager (DCM) provided by the FPGA. The block diagram of clock manager module is shown in Fig. 5.7 (a). The DCM outputs are given to the (i) internal shift controller, (ii) internal process controller, and (iii) enable signal generator.

In these modules, starting from the DCM output, the required enable signals, shift control clocks, process control signals, command signals, data forward signals and baud rate clock etc. were generated at the specified synchronized frequencies for globally controlled data flow and operations.

5.5.2 Signal digitization and data preprocessing unit

This unit consists of two circuits, namely (i) A/D interfacing architecture and (ii) Process Variables (PVs) preprocessor. The A/D interfacing circuit is developed in the FPGA as shown in Fig. 5.7 (b). The data required to enable (En) the analog multiplexer (MSB of MUX o/p) and Channel Select (CS) address (MSB-1: LSB of MUX o/p) are stored in the channel control register. The MUX routes the content of the channel control register in accordance with the present state of 'ch_sel_ctrl' of finite state machine controller. The 12 bit output of A/D corresponding to the selected channel is forced to be available at the input of the DMUX by asserting the Output Enable (OE) control signal. The DMUX input is directed to the respective 12 bit wide register. The selective lines of the MUX and DMUX are synchronized for the proper operation with a finite state machine. Once the content of all the registers are loaded, then the values are forwarded into the PVs pre-processor module. This action is accomplished by the command 'data_sht_ctrl1'

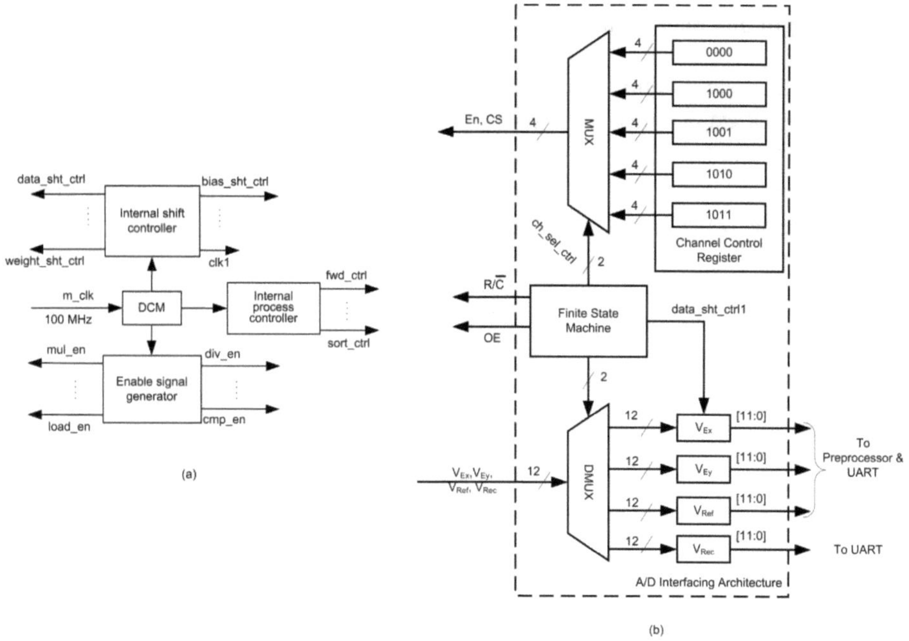

Fig. 5.7: (a) Schematic diagram of digital clock manager unit and (b) eight channel and 12 bit parallel out bidirectional (–10 V to +10 V) A/D interfacing and data acquisition circuit.

from the finite state machine. The preprocessor performs the normalization operation, as given in Eq. (5.6), on the process variables as shown in Fig. 5.3. The radix-10 representation range of digitizer output is 0 through 4095 with the center value of 2048 for the 0 V input and the coding formula is $\sum_0^{11} x_n 2^n$, where x_n is the nth binary digit ($x_n \in [0, 1]$). The V_{Ex} and V_{Ey} are compared with 2048 (0 V) to identify the direction of the beam displacement, i.e. left, right, above or below and the sign bit 'S' is concatenated with the values of V_{Ex} and V_{Ey} with respect to the decision of the comparator. To measure the inner scale magnitude of the beam displacement on OPD, 2048 is subtracted from the values of V_{Ex}, V_{Ey}, and V_{Ref} and then the inner scale magnitudes of V_{Ex} and V_{Ey} are divided by the inner scale magnitude of V_{Ref}, so that V_{Ex} and V_{Ey} are normalized with respect to V_{Ref}. The normalized values of PVs σ_{Ex} and σ_{Ey} are loaded into the data register and forwarded to the D-FFs array. The contents loading and forwarding to/from data register is performed by the control signal 'data_sht_ctrl2'. $Q(t + 1) = Q(t)$ for the rising edge of the 'clk1' and this state is maintained till accomplishing the inputs and synaptic weight multiplications followed by bias addition for the HL1 computations in single precision data format (Dias et al. 2004; Meyer-Bäse et al. 2003; Benrekia & Attari 2010). The next rising edge of 'clk1' is applied to read and forward the subsequent inputs for HL1 computations.

5.5.3 Weight and bias memory management circuit

The synaptic weights and bias values in single precision floating point format are stored in the RAM memory: W^1_{x1}, W^1_{x2} and b^1_x for HL1, W^2_{1x}, W^2_{2x}, $W^2_{3x}, \ldots, W^2_{8x}$, W^2_{9x} and b^2_x for HL2 and W^3_{1x}, W^3_{2x} and b^3_x for OL. The organization of weight and bias memory is shown in Fig. 5.8 and the addresses are allocated based on the priority of data sequence forwarding for successive computations.

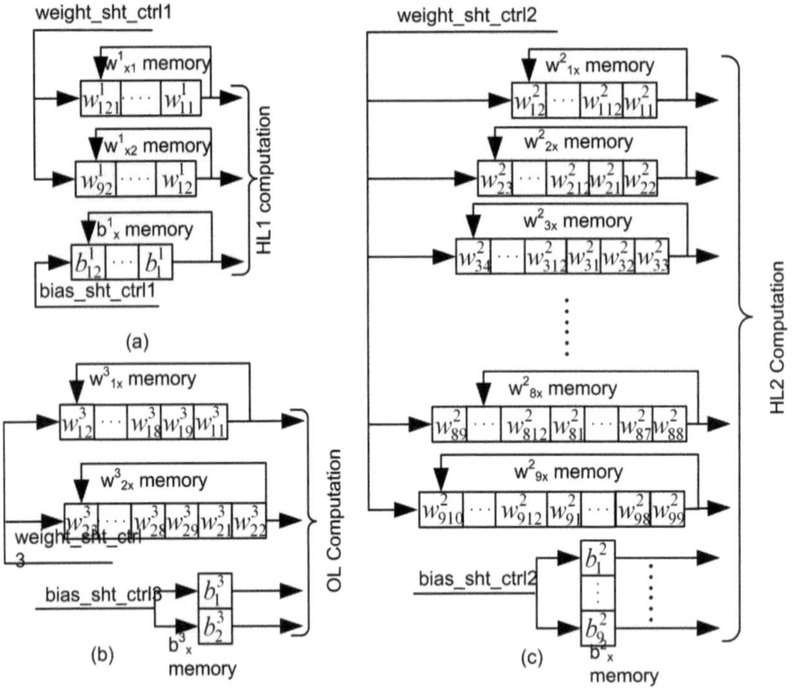

Fig. 5.8: Synaptic weight and bias value memory organization in single precision (32 bit) floating point data format in RAM for (a) hidden layer 1 (HL1), (b) output layer (OL), and (c) hidden layer 2 (HL2).

The RAM units are constructed as a circular data shift register, so that it forwards the chosen weight and bias value to the feedback path as well as to the multiplier (Gomperts et al. 2011). Three 12 element row vectors, nine 12 element row vectors with one 9 element column vector and two 9 element row vectors with one 2 element column vector are the structure of the HL1, HL2 and OL memory. Input, synaptic weight and bias value shifting are synchronized with the clock manager unit and are controlled by the 'weight_sht_ctrl1' and 'bias_sht_ctrl1' signal for HL1, 'weight_sht_ctrl2' and 'bias_sht_ctrl2' signal for HL2 and the 'weight_sht_ctrl3' and 'bias_sht_ctrl3' signal for OL. The weight and bias values are serially shifted to perform the input-weight multiplications, cumulative and bias addition operations to compute the net input of neuron '$n^1(\cdot)$' in HL1. The weights are serially and the biases are shifted in parallel to perform the multiply and accumulation and bias addition operations to compute the neuron net inputs '$n_1^2(\cdot)$ to $n_9^2(\cdot)$' and '$n_1^3(\cdot)$ and $n_2^3(\cdot)$' in HL2 and OL respectively.

5.5.4 Neuron unit

The neuron unit (activation function) is implemented to calculate the activation level of net input 'n' (Tisan & Cirstea 2013). The neurons in the HL1 and HL2 are bipolar sigmoidal activation function and in the OL is a bipolar linear activation function. The bipolar sigmoidal activation function is normally modeled by the hyperbolic tangent function as (Bañuelos-Saucedo et al. 2003; Omondi & Rajapakse 2006; Skoda et al. 2011; Tisan et al. 2009)

$$a = \frac{(1 - e^{-\sigma n})}{(1 + e^{-\sigma n})} = \tanh(\sigma n)\,. \tag{5.8}$$

On differentiating Eq. (5.8), we get

$$a' = \left(\frac{\sigma}{2}\right)[(1 + a)(1 - a)] \tag{5.9}$$

The bipolar sigmoidal activation function cannot be easily implemented in the FPGA since it consists of infinite exponential series as in Eq. (5.8) (Banuelos-Saucedo et al. 2003). The drawback of using a look-up table for bipolar sigmoidal activation function implementation (Skoda et al. 2011) is the need for and utilization of a huge amount of hardware resources, therefore, it is implemented as a second order nonlinear function as in Eq. (5.10) in FPGA (Bañuelos-Saucedo et al. 2003; Biswas et al. 1999) and various other methods of activation function approximation can be found in the literature (Tisan et al. 2009).

$$a = f(n) = \begin{cases} 1 & \text{if } n \geq L \\ n(\beta - \theta n) & \text{if } 0 \leq n < L \\ n(\beta + \theta n) & \text{if } -L \leq n \leq 0 \\ -1 & \text{if } n \leq -L \end{cases} \tag{5.10}$$

where L and –L are upper and lower saturation limits, n is net input to the neuron, β is the slope of the model function and θ is the gain of the model function. The digital architecture of neuron implementation using Xilinx system generator (Xilinx block sets) is shown in Fig. 5.9 (a).

The net input 'n' is multiplied by the gain (θ) and the multiplier output is forwarded to another two processes that add and subtract from the slope (β) and the results are given to the multiplexer which forwards one of two inputs with respect to the selection input (S_1: = 0 first input, = 1 second input). The output of the first multiplexer is multiplied by the net input 'n' and goes to the second multiplexer which

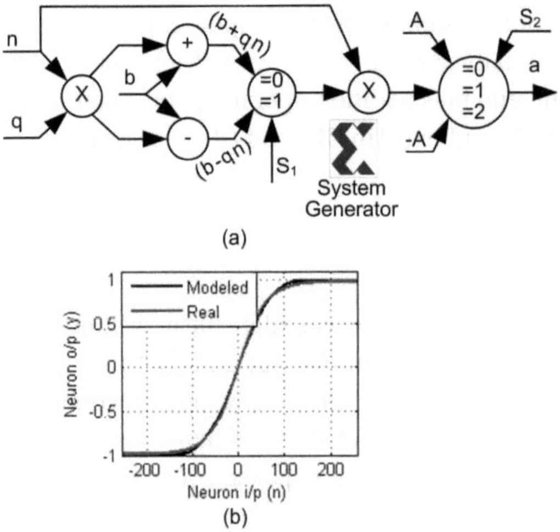

Fig. 5.9: (a) Hardware design flow graph of hyperbolic tangent sigmoidal (bipolar) activation function model as the second order nonlinear function and (b) plot of real and modeled neuron behavior with $\sigma = 0.032$, $\theta = 1$, $\beta = 255$, $n \in \{-255, 255\}$, and $L = \pm\beta/2$.

forward one of the three inputs depending upon the selection input (S_2: = 0 first input, = 1 second input, = 2 third input). The simulation results of real and second order nonlinear bipolar activation function for the input range of $(-255, 255)$ is shown in Fig. 5.9 (b) and the important observations are:

1. The response of the second order nonlinear model precisely fits with the real model
2. The input range, slope, gain, upper and lower limit parameters can be varied based on the application and
3. Possibility of Xilinx block set based effective hardware implementation (Khalil & Al-Kazzaz 2007)

Only one neuron is implemented instead of 12 neurons in the HL1 and full activation level computations are accomplished by the cyclic shifts of inputs, weight and bias (W^1_{x1}, W^1_{x2} and b^1_x), the input and output of the neuron are denoted by '$n^1(\cdot)$' and '$a^1(\cdot)$' respectively. In the HL2, 9 neurons are implemented and the input and output are denoted by $n^2_1(\cdot), n^2_2(\cdot), \ldots, n^2_9(\cdot)$ and $a^2_1(\cdot), a^2_2(\cdot), \ldots, a^2_9(\cdot)$ respectively. The OL consists of 2 neurons and the input and output are denoted by $n^3_1(\cdot), n^3_2(\cdot)$ and $a^3_1(\cdot), a^3_2(\cdot)$ respectively.

5.5.5 Data Routing Ring Circuit (DRRC)

Two Data Routing Ring Circuits (DRRCs) are used to utilize the minimum amount of hardware resources to feed the data to the multiply-accumulation unit as a Serial In Serial Out Circular Shift Register (SISOCSR) for the HL2 and OL configuration as shown in Fig. 5.10 (a) and (b). The DRRC unit is developed based on the methodology of neuron configuration, i.e. one neuron or multiple neurons (Yildirum 2006). Since one neuron is implemented in HL1, a 1×12 demultiplexer is used in the first DRRC to sort the input value into the corresponding second input of a 2×1 multiplexer, i.e. $a^1(\cdot)$ is demultiplexed to $a_1^1(\cdot), a_2^1(\cdot), a_3^1(\cdot), \ldots, a_{12}^1(\cdot)$ as shown in Fig. 5.10 (a) and this sorting is accomplished by the command selection line 'sort_ctrl' signal. Initially, the selection line of the 1×2 multiplexer 'load_en1' is set to '1', therefore, the sorted data are available at the input of the SISOCSR, otherwise the 1×2 multiplexer output depends on the first input, i.e. the output of the previous register. As soon as one set of sorting is over, the register clock 'rot_ctrl1' is applied and then the register's output is available at the first input of 1×2 multiplexers. At this time, the 'load_en1' is set to '0', so that one circular shift is accomplished as $a_{12}^1(\cdot)$, $a_1^1(\cdot)$, $a_2^1(\cdot)$, $a_3^1(\cdot)$, $a_4^1(\cdot)$, $a_5^1(\cdot)$, $a_6^1(\cdot)$, $a_7^1(\cdot)$, $a_8^1(\cdot)$, $a_9^1(\cdot)$, $a_{10}^1(\cdot)$, and $a_{11}^1(\cdot)$. These cyclic rotations are continued for 12 subsequent clock inputs and then repeated with a new set of input data. The first nine registers' outputs are given in parallel to the HL2 multiply-accumulation computations.

The outputs of the HL2 neurons $a_1^2(\cdot), a_2^2(\cdot), a_3^2(\cdot), a_4^2(\cdot), \ldots, a_9^2(\cdot)$ are directly given to the second input of the 1×2 multiplexer of OL DRRC as shown in Fig. 5.10 (b), the

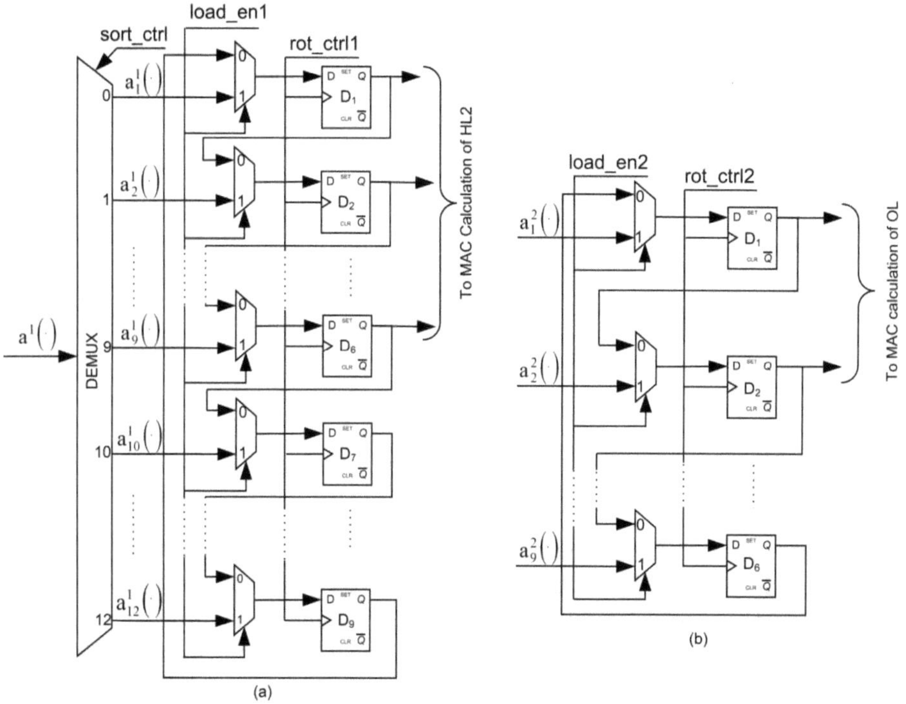

Fig. 5.10: Neuron output data routing ring circuits and serial in and serial out circular shift register array to accomplish the multiply-accumulation computation for (a) Hidden Layer 2 (HL2) and (b) Output Layer (OL).

'load_en2' is set to '1', therefore, the inputs are available at the input of the registers. The register inputs are shifted to the next stage when the clock 'rot_ctrl2' is applied and the status becomes $a_9^2(\cdot)$, $a_1^2(\cdot)$, $a_2^2(\cdot)$, $a_3^2(\cdot)$, . . . , $a_8^2(\cdot)$ and at this time the 'load_en2' is set to '0'. These cyclic rotations are continued for 9 subsequent clock inputs and then repeated with new set of input data. The first two registers' outputs are given in parallel to the OL multiply-accumulation computations.

5.5.6 Multiply-Accumulator unit (MAC)

The two main factors considered during the hardware implementation of Multiply-Accumulator (MAC) units are (i) need for reasonable precision and (ii) cost of more logic area associated with precision (Skoda et al. 2011). The single-precision (32 bit) floating point arithmetic standard is used for accomplishing the input and synaptic weight multiplications with bias additions and all other arithmetic operations (Benrekia & Attari 2010). The two modules of MAC unit architecture are (i) computation of multiplication and (ii) result accumulation. The unit's design using Xilinx system generator (Xilinx block sets) is shown in Fig. 5.11. The nine MAC units for HL2 and two MAC units for OL are developed for neuron inputs $n_1^2(\cdot)$ through $n_9^2(\cdot)$ and $n_1^3(\cdot)$ and $n_2^3(\cdot)$ computations. The first MAC unit section multiplies the outputs of the first DRRC with the corresponding weights and is accumulated.

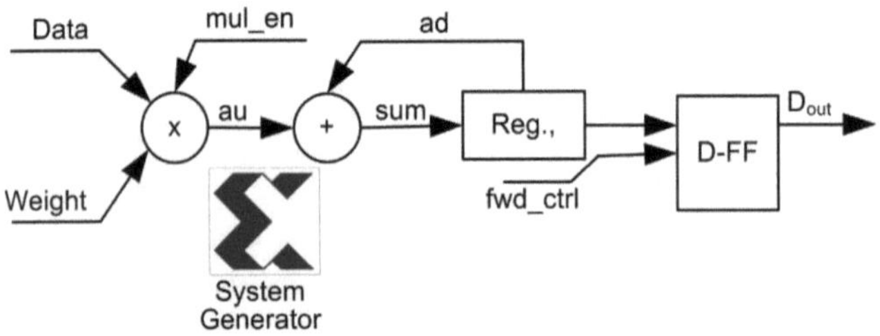

Fig. 5.11: Multiply-accumulation architecture implementation design flow graph.

The weight values and DRRC output shifting are controlled and synchronized by the internal shift controller unit. The command 'mul_en' signal ensures the availability of DRRC outputs and respective weight data and triggers the multiplication process. In accumulation, the multiplier output 'au' is added with the previously accumulated value 'ad' and the result 'sum' is stored in register for future accumulation. Once one MAC operation for one complete set of inputs is accomplished, the final result is forwarded to the bias addition in parallel for the rising edge of command 'fwd_ctrl' signal so that the final output, i.e. the input to the HL2 neurons $n_1^1(\cdot)$ through $n_9^1(\cdot)$, is ready. Similarly, the second MAC unit section multiplies the outputs of the second DRRC with the corresponding weights, accumulates and finally forwards to the bias addition in parallel, so that the OL neural inputs $n_1^3(\cdot)$ and $n_2^3(\cdot)$ are ready. The OL neurons' outputs are given to the D/A circuit.

5.5.7 Serial communication manager

The Universal Asynchronous Receiver Transmitter (UART) RS232 standard communication protocol is implemented as a separate digital architecture in the FPGA as shown in Fig. 5.12 (a). Once data acquisition is complete for one symbol cycle, the values in radix-2 format from the signal digitization unit are directly forwarded to the data buffer in parallel without data conflict/loss. The contents of the buffer are shifted into the communication frame planner for every rising edge of 'data_rd' signal and it splits the measurement data intact length into bytes, concordinates a start bit '0' before every byte and joins sufficient zeros with the bits as shown in Fig. 5.12 (b), to form the communication frame, and then passes on to the MUX1. The output of the MUX1 is forwarded to the Parallel in Serial Out Shift Register (PISOSR) and sends one bit at a time to the MUX2 for the rising edge of 'clk_sig' signal.

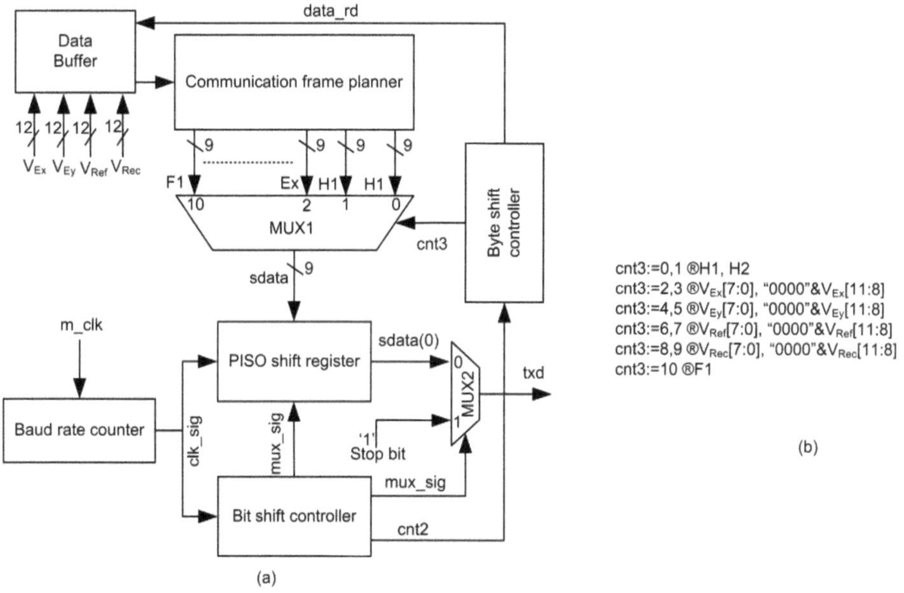

Fig. 5.12: (a) UART RS232 communication protocol architecture and (b) communication frame format.

The output of the MUX2 'txd' is 'sdata(0)' i.e. LSB of PISOSR, when the selection line 'mux_sig' is set to '0', otherwise the stop bit '1'. The baud rate counter generates the bits transmission clock, i.e. baud rate clock 'clk_sig' at the rate of 9600 from the master clock 'm_clk' signal. The transmission of the measurement (M/m) data bit at the PISOSR and the stop bit are controlled by the bit shift controller in accordance with the 'clk_sig'. Further, the bit shift controller conveys the status of the frame transmission to the byte shift controller using 'cnt2', that switches the selection line of MUX1 'cnt3' to choose the next channel (frame). Once all the frames of M/m data corresponding to the one cycle are transferred, then the byte shift controller enables the data buffer to read the new set of M/m data by the command 'data_rd' and continues. In order to avoid the occurrence of communication conflict among the frames, two header (H1 and H2) frames are transferred followed by the M/m data frames and one footer (F1) frame for the appropriate coordination among the data, so that in total 11 frames need to be transferred every data acquisition cycle. A computer (PC1) is used to log all the M/m data to perform the on/off line computations related to the experimental data analysis.

5.6 Experimental results and discussion

The RSM and neural-controller are installed at the FSO link receiver experimental setup as an observer based control for beam stabilization. The performance of the control loop is intensively investigated and the open and closed loop control configuration is characterized with an unmodulated beam. Some of the data acquired experimentally in real time are presented and from them the performance comparisons between these two controllers are carried out. Straightforward experiments are conducted to understand the capability of these two candidate controllers' accuracy, speed and convergence etc. The significance of the proposed controlling scheme in improving the reliability of an FSO link is evaluated using the impulse signal. The outstanding performance of the neural-controller in mitigating beam wandering and reducing detector optical signal fluctuation are analyzed and the results are presented in this section.

5.6.1 Control schemes validation and evaluation in open loop decision-making

The capability of the developed controllers in open loop control configuration, as shown in Fig. 5.5 (b), for mitigating beam wander are tested and the control performances are evaluated. The disturbances are introduced at the transmitter and the corresponding manipulated variables are simultaneously measured at the receiver. The counter position distances are estimated using the measured manipulated variables $C_{1x}(t)$ and $C_{1y}(t)$ from the RSM and $C_{2x}(t)$ and $C_{2y}(t)$ from the neural-controller.

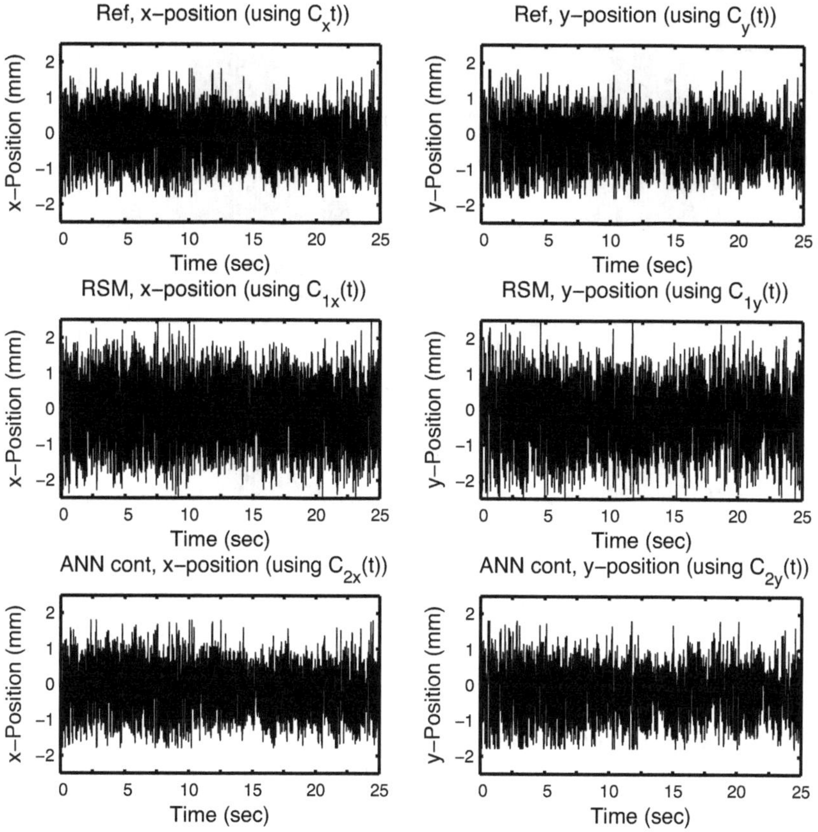

Fig. 5.13: Time series plot of performance of open loop experiment. Upper plot: reference x and y positions estimated using expected control data $C_x(t)$ and $C_y(t)$; middle plot: actual x and y positions estimated using response surface model control data $C_{1x}(t)$ and $C_{1y}(t)$; and bottom plot: actual x and y positions estimated using neural-controller control data $C_{2x}(t)$ and $C_{2y}(t)$.

The time series data obtained in the open loop experiment for 25 s is shown in Fig. 5.13: the upper plot shows the reference (expected) counter position, the middle plot shows the position estimated using RSM controller outputs and the bottom plot shows the position estimated using neural-controller outputs.

The histogram of deviations (errors) of the RSM and the neural-controller control data from the reference data are computed as $e_{1x}(t) = C_x(t) - C_{1x}(t)$ and $e_{1y}(t) = C_y(t) - C_{1y}(t)$ for RSM and $e_{2x}(t) = C_x(t) - C_{2x}(t)$ and $e_{2y}(t) = C_y(t) - C_{2y}(t)$ for the neural-controller and are shown in Fig. 5.14. The error histogram plot corresponding to the RSM in Fig. 5.14 clearly shows that

1. Error distribution falls within ± 1.5 V and follows approximately the trapezoidal distribution
2. More bin frequency is seen from -0.9 V to 0.8 V

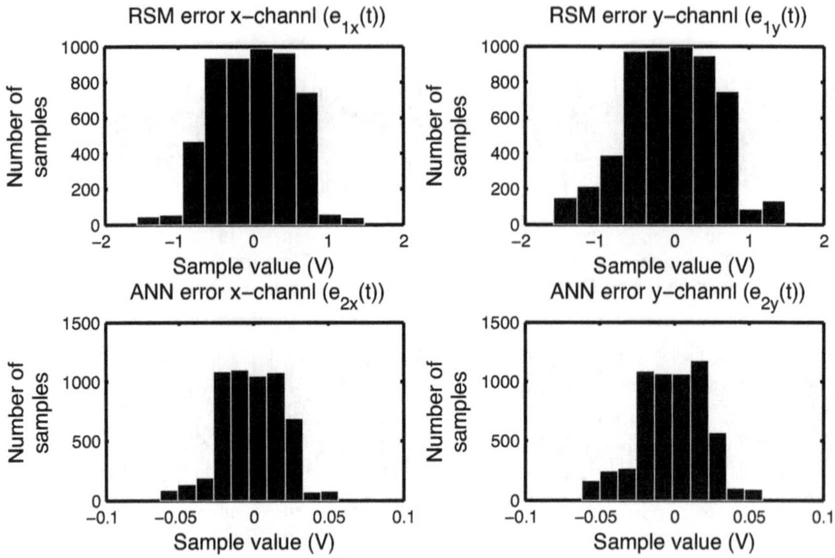

Fig. 5.14: Histogram of time series plot of control data deviations (errors). Upper plot: deviation errors $e_{1x}(t)$ and $e_{1y}(t)$ estimated from response surface model. Bottom plot: deviation errors $e_{2x}(t)$ and $e_{2y}(t)$ estimated from neural-controller.

3. RSM generates phase shifted control data
4. RSM fails to keep good correlation with the reference value
5. RSM maintains nonlinearity in the deviation errors and
6. High control data repeatability around 0 mm

The bottom plot in Fig. 5.14 for the neural-controller shows that the predicted output is close to the reference values with the deviation error band of –0.07 V to +0.06 V. The sample count is more in the range of –0.02 V to +0.02 V. Further, the neural-controller does not yield any phase shifted output even at high frequency. Thus, better enhancement in the control surface response is achieved using the neural-controller.

Tab. 5.2: Validation test results of developed response surface model (full) and neural-controller for C_x and C_y.

Trial No.	Response Surface Model				Neural-controller			
	Prediction value (V)		Prediction error (%)		Prediction value (V)		Prediction error (%)	
	C_x	C_y	C_x	C_y	C_x	C_y	C_x	C_y
1	1.26	3.43	8.97	–9.25	1.37	3.11	0.00	0.01
2	–2.24	–5.12	2.53	–2.35	–2.29	–5.00	0.02	–0.02
3	6.00	–4.90	4.68	–8.16	6.28	–4.50	0.01	–0.01

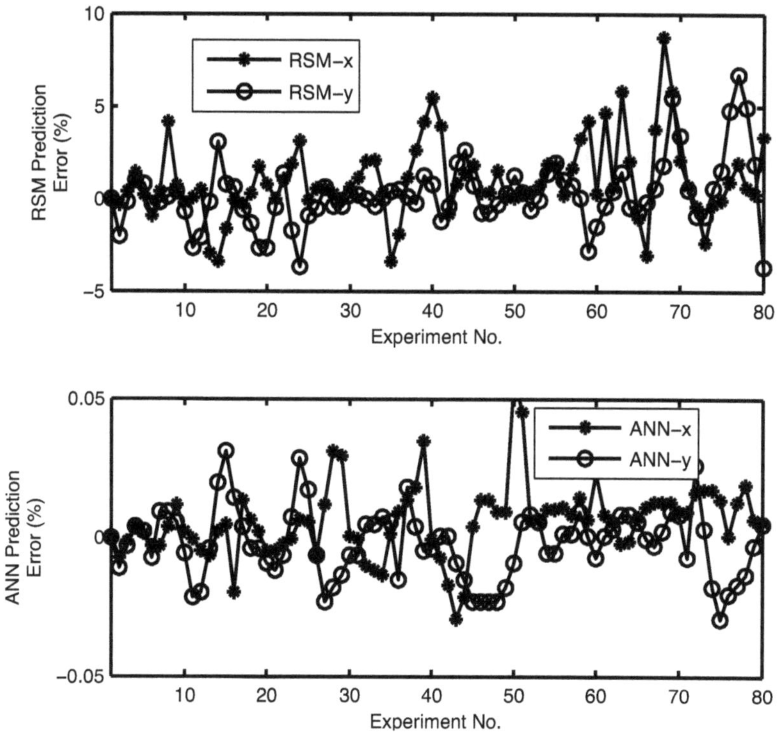

Fig. 5.15: Error percentage of control signal predicted by the response surface model and neural-controller in open loop configuration for x- and y-channel.

The prediction values of the RSM and neural-controller for the new values of process variables under validation test are measured and the correlation level with the observed data are estimated in terms of prediction error percentage using

$$\frac{\text{Observed value} - \text{Prediction value}}{\text{Prediction value}} \cdot 100$$

as given in Tab. 5.2 and the estimated prediction error percentage for the pair of 81 new input factors in the open loop test from RSM and the neural-controller are shown in Fig. 5.15.

The percentage of prediction errors obtained from the RSM varies in the outer scale with the min/max values of −4.88 % and 9.21 %, whereas the neural-controller is almost very close to 0 % with the min/max values of −0.03 % and 0.05 %. The neural-controller takes almost close corrective action so as to bring back the errors to (0, 0). Thus the neural-controller predicts the exact control surface. This control performance clearly evidences that the RSM is deficient for real-time control applications wherever the prediction accuracy is significant.

The neural-controller has a good capability to track and stabilize the beam to a particular point. Furthermore, good beam stabilization is obtained even for the z_{offset} changes. Therefore, it is concluded that the neural-controller is the better choice over the RSM, it is adequate and more suitable for coupling the PIB into the detector plane, especially in FSO communication. Therefore, the neural-controller alone is retained at the receiver and the subsequent data analyses are carried out.

5.6.2 Performance study of closed loop experiment with intensity feedback control

The maximum and constant performance of the controller could not be arrived at by open loop control configuration, since the control in open loop depends on

1. How accurate the estimation of the position errors using OPD is
2. The long-term behavior of optoelectronic components
3. Number of piezo-actuator motion cycles
4. Magnitude and phase response of the FSM,
5. Sensor noise and
6. Initial LoS alignment

Therefore, closed loop control configuration becomes significant. In the system considered in this work, the optimum values are $C_x^* = 0.2157\,V$ and $C_y^* = -0.1634\,V$ and these two values are very important in the experiments. As a direct consequence of the disturbance rejection in each channel, the average value of the position errors are significantly closer to (0, 0) mm. The convergence test experiments to analyze the response of the neural-controller for the disturbances in closed loop control configuration as in Fig. 5.5 (c) are conducted and samples of the results are shown in Fig. 5.16. It can be seen that the greatest performance improvements are attained with the neural-controller in closed loop control configuration.

The time series data are evaluated using the empirical standard deviation denoted by σ_x, σ_y and position magnitude average denoted by β_x, β_y (MATLAB) in open and closed loop control configuration for the x- and y-channel respectively for a longer period as given in Tab. 5.3.

These statistical results confirm that the neural-controller in closed loop generates the accurate control signal as required for mitigating the disturbance. The Power Spectral Density (PSD) is computed for the time series data to analyze the disturbance rejection as a function of frequency and is shown in Fig. 5.17. The disturbance spectrum given to the x-channel has the low frequency content of 0 through 120 Hz with three narrow bands while the y-channel has a disturbance spectrum composed of 0 through 340 Hz with three narrow bands. The correction input to the FSM has a frequency of 1.43 kHz. The PSD results show that the proposed control scheme is capable of rejecting disturbances in the range 0 through 1.43 kHz or higher up to 2 kHz.

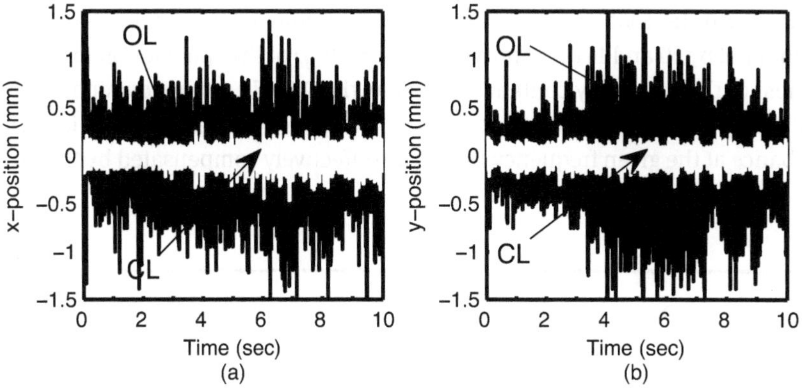

Fig. 5.16: Time series plot of beam displacements on the OPD in open and closed loop control configuration for (a) x- and (b) y-position.

Tab. 5.3: Empirical standard deviations and position magnitude average of experiments using the population length of 10 000 data points in open and closed loop control configuration.

Trial ID	Normalized error standard deviations			Position magnitude average (mm)		
	Estimation	**Open loop**	**Closed loop**	**Estimation**	**Open loop**	**Closed loop**
1	σ_{x1}	0.68	0.17	β_{x1}	0.53	−0.01
	σ_{y1}	0.66	0.13	β_{y1}	0.42	0.13
2	σ_{x2}	0.73	0.18	β_{x2}	0.38	−0.01
	σ_{y2}	0.77	0.15	β_{y2}	0.68	0.14
3	σ_{x3}	0.79	0.24	β_{x3}	−0.32	−0.15
	σ_{y3}	0.78	0.21	β_{y3}	0.37	0.15

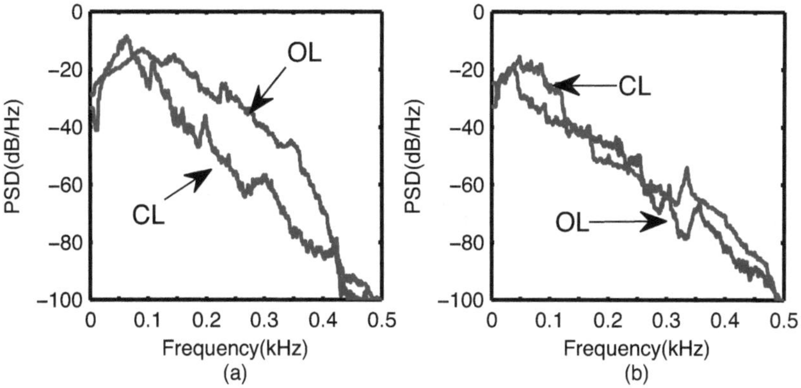

Fig. 5.17: Comparison of Power Spectral Density (PSD) of time series of position data in open and closed loop control configuration for (a) x- and (b) y-channels.

The PSD demonstrates that the disturbance band is attenuated in x-channel by 30 dB and in y-channel by 10 and 15 dB approximately. The PSD of the spot motion indicates that the most of the frequency content falls below the 0.5 kHz as already reported by Suite et al. 2004. Almost the same observations are reported for a longer period and the disturbance at the given frequency ranges are effectively compensated by the proposed control system.

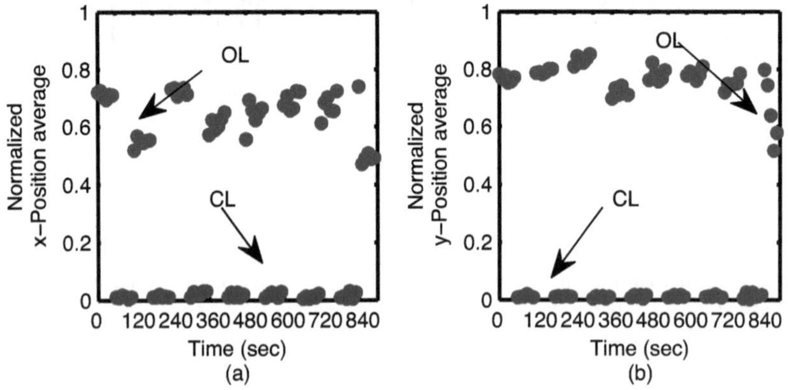

Fig. 5.18: Alternative (open (red) and closed (blue) loop control) 10 second average values of absolute values of position displacement on (a) x-channel and (b) y-channel for 900 seconds in a 1 minute interval.

Further, the results show that the neural-controller is capable of significantly attenuating the disturbance affecting both channels. In this case, what is interesting to note is that the overlapping frequency bands are noticeably attenuated, however, not as much as in the cases in which the frequency bands do not overlap due to cross interference between both channels as commented in the literature (Perez-Arancibia et al. 2012; Suite et al. 2004). The average values of the absolute values of x- and y-channel position data corresponding to open and closed loop control configurations are calculated in the ten second intervals. The alternative sample average data set calculated for the 900 s in one minute interval is shown in Fig. 5.18. The average value changes are seen almost within 0.40 and 0.75 in open loop, whereas in closed loop they are 0 to 0.0021 which directly means that the best beam stabilization is achieved and a good agreement with the experimental data reported in Perez-Arancibia et al. (2012) and Trisno et al. (2004) is kept.

5.6.3 Analysis of beam spot auto alignment and reduction of wandering

Optical beam angular deflections along the propagation path wander the beam randomly out-off LoS between the optical source and the detector which severely affects the coupling of PIB into the detector when the angle of arrival fluctuation is not compensated by the controller. The instantaneous center of the beam, i.e. the point of maximum intensity (since focused Gaussian beam), is randomly moving on the detector plane due to this effect. The trend of angle of arrival of optical beam in both axes at the detector plane without any control is estimated using Eq. (5.5) and a portion of the data collected on 11th January 2014 is shown in Fig. 5.19. The main observation from Fig. 5.19 is that the standard deviation changes significantly which indicates that an effective beam tracking and positioning is important and subjected into the compensation control process. The performance improvement in beam stabilization through the angle of arrival compensation control system is tested in controller turned-on and turned-off conditions. The beam centroid coordination on the 2D plane of OPD is estimated using Eq. (5.3) and a portion of data are shown in Fig. 5.20 (a): with compensation control off (blue) and on (red).

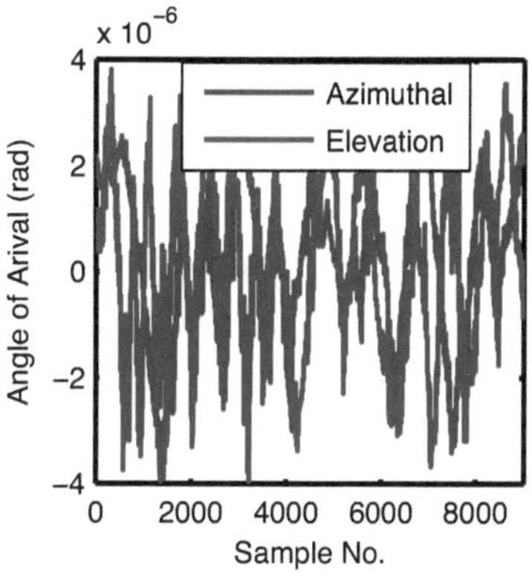

Fig. 5.19: A portion of beam angle of arrival measurement carried out on 11th January 2014 when z_{offset} is constant.

The interesting observations from Fig. 5.20 (a) in no control are
1. The beam centroid wanders almost all the area of the detector plane
2. The centroid movements are highly random
3. Sometimes the centroid goes out-off the detector plane, i.e. the FSO link is disconnected and
4. A wide range of power fluctuation at the communication detector is seen

In control are
1. Centroid movements are controlled and are always around the optimal point
2. A very narrow range of power fluctuation are seen and

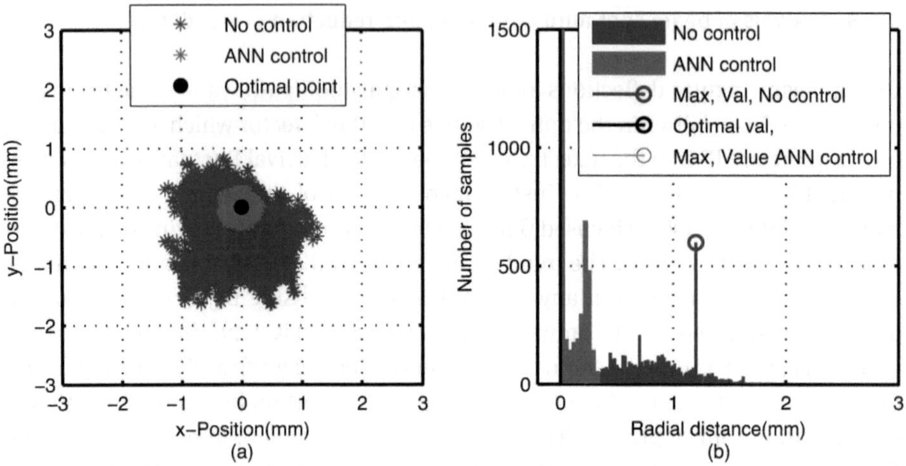

Fig. 5.20: Laser beam spot centroid motion with and without beam AoA compensation control: (a) beam wandering on 2D-4Q plane with horizontal and vertical scales of real space. (b) Histogram of plane center to beam centroid hypotenuse distance.

3. Maximum intensity of beam (Gaussian beam) is concentrated at the center of the communication detector.

The radial distance of a line connecting the center of the OPD and beam centroid is calculated using Eq. (5.4) in no control (blue) and in control (red) conditions and their corresponding histogram plot is shown in Fig. 5.20 (b). The plot evidences that the beam angle of arrival compensation control system mitigates beam wandering and stabilizes the beam almost at the center of the communication detector.

5.6.4 Behavioral study of effective scintillation index and impulse response

The (0, 0) V output from the OPD means that the beam centroid is positioned on the center of the photodiode (communication detector) and its output is exactly the minimum possible value, i.e. it should coincide with the lower bound of the photodiode output $V^*_{Rec} = 0.3162$ V. In the absence of sensor noise, the photodiode output difference ($V_{Rec}(t) \sim V^*_{Rec}$) measured in volts is directly related to beam scintillation (power fluctuation due to the atmospheric turbulence effects) and effective scintillation (power fluctuation resulting from the nonperfect coupling of the PIB into the communication detector due to beam wandering). The scintillation effects (short-term data losses) can be recovered using various data coding techniques developed for wire and fiber-based communication systems. In contrast, the effective scintillation (beam wandering induced deep signal fading) represents a unique and significantly more challenging problem, which cannot be solved using conventional data coding techniques without incorporating compensation controls with an optoelectronic assembly (Majumdar & Ricklin 2008). This effect is mitigated using the compensation control system, so that beam stability on the photodiode is improved. The power fluctuations of the received signal are evaluated with and without compensation control. The ensemble average of optical intensity measured in terms of voltage is used to estimate the effective scintillation index using $\{(\langle V_{Rec}^2\rangle / \langle V_{Rec}\rangle^2) - 1\}$ and is shown in Fig. 5.21.

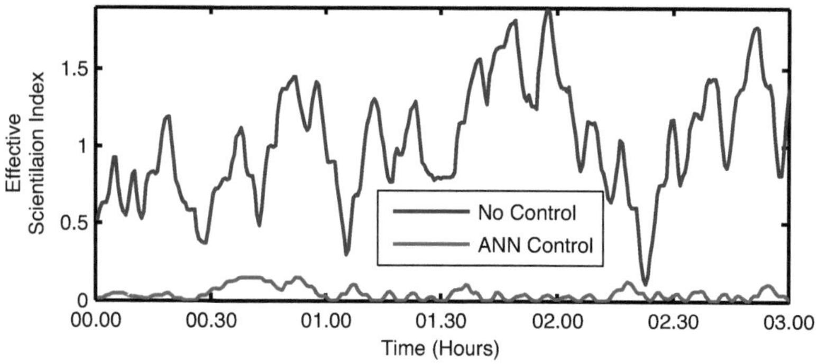

Fig. 5.21: Time series plot of effective scintillation index measured at the receiver station (after propagating 0.5 km) using ensemble average of irradiance of optical signal with beam wandering compensation control turned on (red) and off (blue).

It can be clearly seen from Fig. 5.21 that the effective scintillation parameter fluctuates within the inner scale with the min/max values of 0 and 0.29, i.e. 0 % to 21.17 %, when the beam compensation is turned-on, and the broader variations from 0.19 to 1.36 (outer scale), i.e. 13.87 % to 97.65 %, are seen when the compensation is turned-off. This result exhibits the efficiency of the neural-controller in mitigating beam wandering as required. Therefore, the neural-controller in closed loop control system improves the reliability of an FSO communication data link. The effective scintillation parameter fluctuation, even with beam steering turned-on condition, reaches the maximum value of 0.2964. This is purely due to the scintillation optical power fluctuation in the transmission path (refractive index of channel, Mie/Rayleigh scattering, absorption, and particle interactions with the optical beam etc.) which could not be controlled by the steering system and is a matter for separate research. As a direct consequence of the disturbance rejection in each channel, the average value of the photodiode output is significantly closer to the optimal value V_{Rec}^*, despite the fact that $\{\sigma_x(t), \sigma_y(t)\}$ and $\{C_x(t), C_y(t)\}$, $\forall t$.

All the above analyses are conducted with an unmodulated optical beam. Now the sequence of impulse (UP-NRZ) in a longer interval is given to the laser modulator at the transmitter and the response of the receiver (after propagating 0.5 km) is studied with and without beam compensation control as performed in Safari & Uysal 2008.

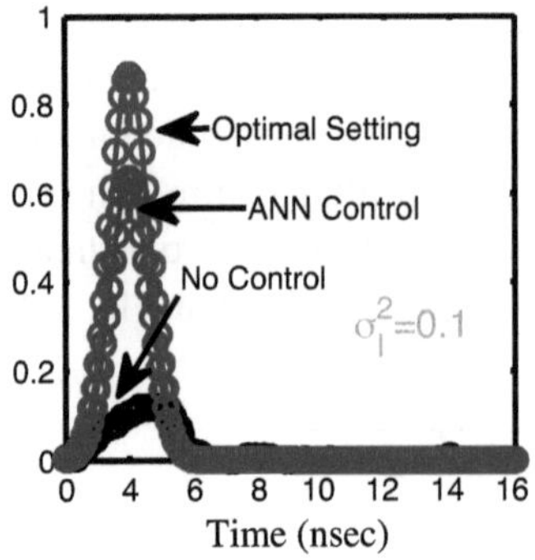

Fig. 5.22: Normalized impulse response of samples average verses impulse time with and without beam steering.

The normalized samples' average values of impulse signal measured in volt at the receiver for the optimal LoS setting, beam control tuned-off and tuned-on conditions are shown in Fig. 5.22 and it clearly shows that

1. The sample magnitudes reach 0.924 in the optimal setting due to less transmission loss in clear atmosphere ($\sigma_I^2 = 0.1$) and 0 mm beam wandering
2. The sample magnitudes do not exceed 0.152 in the absence of beam wandering compensation
3. The received signal magnitudes are significantly improved and close to the optimal magnitudes in the presence of beam wandering compensation and increase the reliability of the FSO link
4. A good stability is maintained at the receiver power and SNR which resulted in decreased BER and
5. Average magnitude for the 0 V input is 0.0372 V due to sensors and circuit noise.

5.7 Summary

The establishment of a terrestrial FSO link for the range of 0.5 km at an altitude of 15.25 m is explained. The RSM controller formulated based on Taguchi's principles is implemented in FPGA and the individual and combined effects of the input factors on the response are analyzed using the ANOVA study. The optimal weights and biases (goal $\leq 10^{-6}$) are obtained from the MATLAB environment when a 2–12–9–2 structure neural network of multilayer perceptron model is used for hardware design. Novel digital architecture designed for the hardware implementation of a neural-controller in the FPGA is discussed in detail. The developed architecture for UART RS232 standard communication protocol is used to transfer (store) the measurement/computational data to the computer and subsequently on/off line plot profile and data analysis are carried out. The validation test for RSM and neural-controllers in open loop decision-making are conducted and the min/max value of exactness deviation of ±1.5 V from an RSM controller and approximately ±0.06 V from a neural-controller are achieved.

The maximum percentage of prediction error of −4.88 % and 9.21 % for RSM controller and −0.03 % and 0.05 % for neural-controller are obtained from the confirmatory tests. A greatest control response is achieved from the neural controller in closed loop control configuration. The capabilities of mitigating the power fluctuation are evaluated with and without beam wandering compensation technique using unmodulated beam. The values of the effective scintillation index are estimated and brought down to the controlled range of 0 % to 21.17 %. The impulse response of the entire system is practically studied with and without compensation control. The results presented in this paper are intended for setting up the FSO communication transceiver. So it is suggested that the use of neural controller for fine steering applications is best instead of complex/complicated computational algorithms found in literatures. The proposed neural controller can also be used in various applications including beam stabilization, steering, focusing, beam spotting, beacons beam coupling and real time control system (Bazil Raj et al. 2014d, 2015c, with permission of Springer).

6 Quality metrics and reliability analysis of ground-to-ground free space laser communication in different weather conditions together with beam steering system

Over the past two decades Free Space Optical Communication (FSOC) has become more and more interesting as an adjunct or alternative to Radio Frequency (RF) communication. Optical wave propagation in an FSOC channel is severely affected by turbulent atmospheric parameters and they lead to the degradation of the data transmission quality and reliability. Among the various atmospheric effects, beam wandering is the main cause for major power loss and cannot be solved without incorporation of a Beam Wandering Compensation (BWC) control system. Therefore, a closed form treatment describing beam wandering effect mitigation is required. Simplex FSOC data transmission (data rate is 155 Mbps) transmitter-receiver experimental setups are established for the data link range of 0.5 km at an altitude of 15.25 m at the college campus. A near IR 850 nm optical source is used for data transmission. The more accurate values of atmospheric attenuation (Atten) and turbulence strength (C_n^2) are obtained using the new models developed according to local weather data. The transmittance of the FSOC channel is calculated using the estimated atmospheric attenuation. The methods of measuring the Q-factor and Bit Error Rate (BER) by postprocessing the received signal are described and used to validate the performance of the FSOC system. The improvements on received signal strength and centroid stability by incorporating the BWC control system are analyzed with various communication key parameters in the very weak to very strong atmospheric turbulence conditions. The average BER at a value of $9.82 \cdot 10^{-9}$ is maintained with the Q-factor value of ≈ 6 in almost all atmospheric turbulence conditions with compensating beam wandering effects, i.e. pointing error losses. Evidence of the suitability and effectiveness of the developed controller in terms of received power (P_r), transmittance (T), eye diagram, Q-factor, Signal to Noise Ratio (SNR), BER, link margin and decoding decision threshold values are comprehensively analyzed through experimental results and reported in this chapter.

6.1 Introduction

The features of Free Space Optical Communication (FSOC) technology make it more viable compared with other existing Radio Frequency (RF), microwave and millimeter wave based wireless communication technologies. FSOC is becoming an increasingly essential part of modern information technology (Ferdinandov et al. 2007). This is due to a number of key advantages including a large unregulated (free of government regulation) and license-free transmission bandwidth spectrum, large data transmission, low power consumption, security (cannot penetrate walls) as well as immunity to Electro-Magnetic Interference (EMI) and Electro-Magnetic Compatibility (EMC) (Afonso et al. 2006; Perez et al. 2014; Cowley et al. 2011). FSOC connectivity can effectively be utilized to process a digital data signal coming from a standard transmission line by exploiting the inherent parallelism of optics (Boffi et al. 2000). FSOC will play an important role in future satellite networks and space scientific exploration that include (i) satellite crosslinks, (ii) satellite to aircraft, (iii) aircraft to aircraft, and (iv) ground station to aircraft and/or satellite and vice versa (Wan et al. 2010). The portability of FSOC is leading to the establishment of temporary links for big and high speed information flows in cities and industrial complexes (Ferdinandov et al. 2007). In Metropolitan Area Networks (MANs), FSOC can be used to enable Local Area Networks (LANs) to LAN connectivity and intercampus connectivity, since FSOC is an excellent candidate for last mile connectivity (Djordjevic 2010). The data rate in RF is limited to the end users (Leitgeb et al. 2009). The available data rates in Unregulated 2.4 GHz Industrial, Scientific and Medical (ISM) wireless LANs are 1–2 Mbps (Sikora & Groza 2005), 20 Mbps at 5.7 GHz and 875 Mbps in 60 GHz (Rakotondrainibe et al. 2009). In order to support the high bandwidth services, the use of microwave or millimeter (mm)-wave systems will be required. But, in these systems, device technology is currently either expensive or immature. Further, the RF spectrum is already exceedingly congested and frequency allocations of sufficient bandwidth are extremely hard to obtain (Singh et al. 2004). The main differences between RF and FSOC technology are summarized and given in Tab. 6.1.

A comparison of FSOC with various other wireless technologies and fiber is shown in Fig. 6.1 and it exhibits that FSOC offers fiber-like data rates up to a link range of 10 km with advanced optical source, detector and optoelectronics devices. As an alternative to RF technology, FSOC systems offer a large data bandwidth up to 2000 THz making them a very attractive technology for meeting future demand for broadband

Tab. 6.1: Comparison between RF and FSOC communication systems.

Prime parameters	**RF**	**FSOC**
Data rate	< 100 Mbps	10 s Mbps–10 s Gbps
Devices size	Large	Small
Bandwidth regulated	Required	Free license
Security	Low	High
Network architecture	Nonscalable	Scalable
Link performance effects	Multipath fading, rain, interferences	Fog, atmospheric turbulence, misalignment, and obstruction
Channel distance	Long	So far small
Noise	Other sources	Background light
Fog Attenuation (V = 200 m)	3 dB/km at 58 GHz	37 dB/km at 830 nm

internet access and High-Definition TeleVision (HDTV) broadcasting services (Sinefeld et al. 2013; Saari et al. 2013). FSOC with visible optical spectrum is emerging as a key technology for a ubiquitous communication system, because the LED has the advantages of fast switching, long life expectancy, being less expensive and safe for the human body (Kumar et al. 2008). FSOC does not incur high costs for digging up tunnels and trenching of roads to install bulky copper and fiber cables (Matsumoto et al. 2008). Therefore, the installation and maintenance cost of an FSOC link is much lower than RF and fiber-based links as shown in Fig. 6.2. These characteristics lead to large interest in the wide and versatile scientific and engineering development of FSOC components with suitable means of atmospheric effect compensation for reliable installation of FSOC data link.

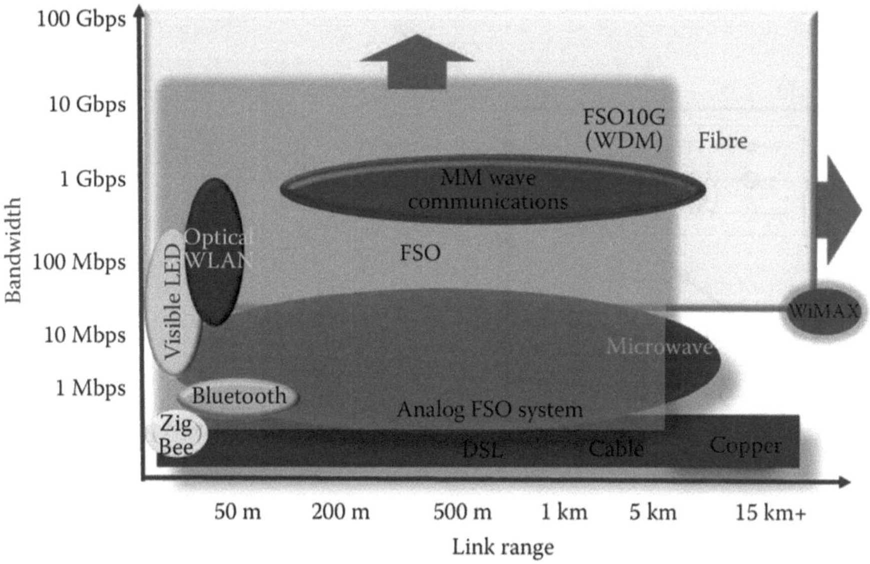

Fig. 6.1: Comparison of range of RF, fiber, and FSOC in terms of available data rates against link ranges (Ghassemlooy et al. 2012a).

Despite these advantages, the performance of FSOC links is largely dependent on weather conditions and their development is related to several key problems (Mahdieh & Pournoury 2010; Ijaz et al. 2011). They result from multiple physical conditions: fluctuations of atmospheric extinction (atmospheric transmission) and fluctuation of the propagation direction of the beam resulting from mechanical vibrations of various origins of the transmitting antenna and from atmospheric turbulence during transmission that affect the temporal and spatial structure of the transmitting optical wave (Zilberman et al. 2010; Kazem 2013; Zhao et al. 2010a). These are the general problems that require attention to be dealt with. The major impact on the FSOC data link due to random fluctuations of atmospheric turbulence is as follows: (i) optical scattering and observation (characterized by atmospheric attenuation), (ii) random optical power fluctuation (characterized by scintillation index), (iii) beam wandering, i.e. beam surface global (temporal) tilt due to beam centroid displacement on the detector plane (characterized by effective scintillation index), and (iv) wavefront distortions i.e. beam surface (spatial) local tilt (characterized by the Zernike polynomials) (Arnon & Kopeika 1997; Liu et al. 2014). The conventional data coding and/or modulation techniques (Djordjevic 2007) can effectively be used to equalize/compromise the first two effects with respect to weather conditions at a given instant of time (Djordjevic 2010; Leibowitz et al. 2005). The third and fourth effects cannot be compromised without incorporating adaptive optics elements (Arnon & Kopeika 1997; Zocchi 2005; Pedireddi & Srinivasan 2010; Zhao et al. 2010b; Liu et al. 2014). Resolving these problems is necessary for FSOC research and design (Ferdinandov et al. 2007).

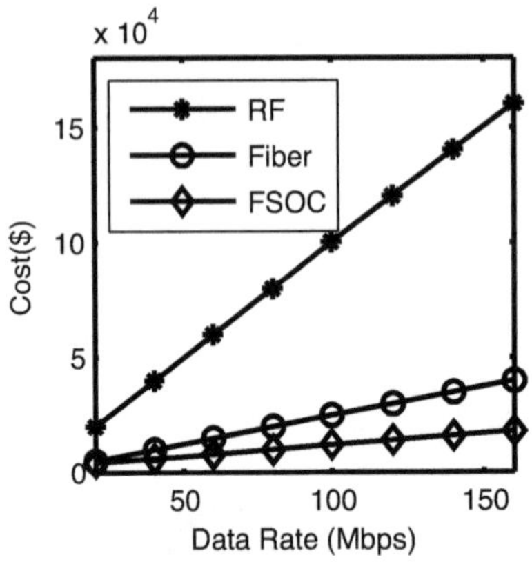

Fig. 6.2: Comparison of bandwidth cost for RF, fiber and FSO based communication system (Willebrand & Ghuman 2002).

Optical beam propagation through the turbulent atmospheric channel suffers from attenuation and phase distortion in horizontal propagation rather than vertical (Djordjevic 2010; Perez et al. 2014). All these effects result in both average received power loss and instantaneously received power fluctuation, called fading, at the receiver aperture and lead to increased system BER and decreased channel capacity and reliability (Zhao et al. 2010a). These effects interrupt the high data rate of an FSOC link from achieving the availability of five nines 99.999 %. One approach to address this problem is increasing the optical power level launched into the air when the propagation loss becomes high. However, the use of high-power Optical Fiber Amplifiers (OFAs) as boosters will increase the cost steeply (Hurh et al. 2005). Another approach to mitigate the power loss and fading problems is to enhance the FSOC channel in the optical propagation domain, for example, multiple aperture transmitter and multiple aperture receiver (Carruthers & Kahn 2000), Adaptive Optics (AO) (Zhao et al. 2010b), and larger receiver aperture sizes (Fath & Haas 2013; Pedireddi & Srinivasan 2010). For

beam wander induced optical power fading, a slightly diverged beam may be used to reduce deep fades. However, a diverged beam usually causes significant loss of average received power. If the receiver aperture diameter is larger than the coherent radius, a larger aperture can reduce the fading level. While larger receiver aperture generally collects more optical signal with reduced channel fading, it also collects more background optical noise (Leibowitz et al. 2005). Therefore, the benefits of using slightly diverging beam and/or larger receiver aperture (larger than or close to the long-term laser spot size) may not balance FSOC performance loss due to pointing misalignment (Liu et al. 2014; Pedireddi & Srinivasan 2010). Therefore, Wireless Optical Communication (WOC) is still considered an emerging technology and is widely being researched, particularly with regard to mitigation of signal degradation caused by propagation through a turbulent atmosphere (Kedar & Shlomi 2005).
Further, understanding the real-world open atmospheric effects to the data bearing optical wave at the installation field is important to characterize the maximum data rate that the system could operate at and to develop atmospheric effects compensation systems. Developing a long range FSOC data link and using it for experimentally studying atmospheric effects and testing and validating the characteristics of the systems developed for compensating the atmospheric effects becomes significant, but it leads to a huge amount of expenses and consumes much time. Therefore, a 0.5 km FSOC data link is established as shown in Fig. 6.3 and used to conduct the on-ground experimental study to understand channel behavior, effectiveness of the system developed for compensating the beam wandering effect and improvements on communication parameters.

The rest of the chapter is organized as follows: Section 6.2 presents the background and related works, Section 6.3 discusses the theory and numerical methods

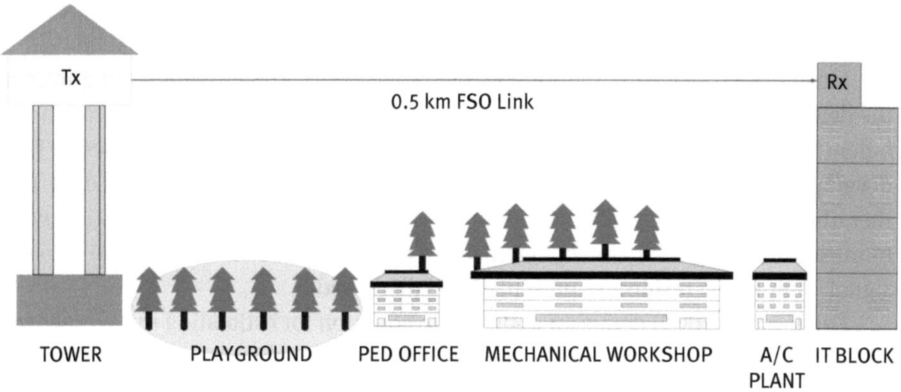

Fig. 6.3: Overall layout of constructed free space optical communication data link with the illustration of buildings and playground in between Tower (Tx) and college Information Technology (IT) block (Rx).

used to quantitatively estimate and measure atmospheric attenuation, turbulence strength and communication key parameters, Section 6.4 describes the experimental setup constructed for a 0.5 km FSOC data link, beam spot position analysis and processing of received data, Section 6.5 discusses the experimental results and data analysis, and finally Section 6.6 draws some conclusions.

6.2 Background and related works

In the last few years, many simulations with various stochastic statistical assumptions of turbulence strength and field measurements with controlled atmospheric conditions (using a turbulence simulation box inside the laboratory) related to the analysis of the quality and reliability of FSOC have been carried out and can be found in the literatures. The publications/results closely related to our work are discussed in this section.

Arnon and Kopeika (1997) derive a mathematical model of an adaptive communication system with a multiscattering channel. With the knowledge of the impulse response function of the cloud, the communication parameters are adjusted to the transfer function of the cloud. Transmitter parameters are varied with respect to the atmospheric conditions and the bit rate is changed accordingly. Also they describe the presence of cloud between the transmitter and receiver that causes temporal broadening and attenuation of received signal.

Daniel L. Hutt and David H. Tofsted (2000) investigate the effect of atmospheric optical turbulence on Ultra Violet (UV) radiation with a wavelength of 253.7 nm. In the experiment, the normalized irradiance variance was measured using a UV scintillometer with a path length of 185 m. Atmospheric turbulence strength is estimated and simultaneously measured. Received signal quality is continuously measured and characterized with respect to different aperture averaged values.

Jacob C. Brandenburg and John Q. Liu (2009) study the probability of a miss in the detection of a signal in an intensity modulation based FSOC system through turbulent atmosphere. The channel behavior is assumed to be a log-normal process. They develop various mathematical methods to analysis and decode the received data. Further, they describe a way for determining the probability of missing an optical communication through the atmosphere using a characteristics function of the received signal. Simulation results are presented and discussed.

Katsis et al. (2009) derive an appropriate equation for estimating the average (ergodic) capacity and the Bit Error Rate (BER) of a log-normal FSOC channel in the case of weak to moderate atmospheric conditions using the frequentist and Bayesian approach. The derived analytical expressions are verified by various numerical examples and simulation results and the data are discussed.

Latsa Babu Pedireddi and Balaji Srinivasan (2010) established a 50 m folded free space optical path and discuss the various effects of atmosphere turbulence such as scintillation and beam wandering. They characterize the effects of inner and outer scale turbulence strength using the 50 m FSOC data link. Discrete Wavelet Transform (DWT) based signal processing is used to increase signal detection accuracy and reduce BER.

Mahdieh and Pournoury (2010) present the numerical evaluation of BER by studying the propagation of an initial Gaussian laser beam through turbulent atmosphere. Beam scintillation, spreading, and wandering are considered as atmospheric turbulence effects on laser beam degradation. The simulation results for different conditions are presented and analyzed.

Popoola et al. (2010) describe the impact of atmospheric turbulence: induced fading on the symbol decision position in OOK and Binary Phase Shift Keying (BPSK) Subcarrier Intensity Modulation (SIM) optical communication link. The different turbulence conditions are simulated in the laboratory using a chamber equipped with heating elements and fans. BER against electrical SNR are estimated for OOK and BPSK-SIM schemes in various artificial (using chamber) atmospheric conditions.

Fabio E. Zocchi (2005) develops various analytical models to analyze the performance of an adaptive optics system for FSOC direct detection system. The author calculates the transfer function of the adaptive control loop in the Zernike and Fourier domains for temporal and spatial dependence. The Power Spectral Density (PSD) of the corrected wavefront is determined. Simulation results are analyzed and he concludes that the model can further be upgraded to refine its predicting capability.

Zabih Ghassemlooy et al. (2012b) investigate the performance of FSOC using a manmade atmospheric turbulence chamber at the data rate of Ethernet (12.5 Mbps) and fast-Ethernet (125 Mbps) under different artificial turbulence conditions. The Q-factor and BER quantitative analysis are carried out and the results are discussed. The dependence of the communication parameters with the FSOC turbulent channel is analyzed. The indoor analyses are related to the outdoor FSO link using a simple function.

Wei Liu et al. (2014) established a few meter FSOC data link using the developed turbulence simulation box in the indoor laboratory environment and estimate the BER with open and closed loop beam steering control systems. The experimental results showing improvements on BER with the closed loop beam steering control system are highlighted with the experimental results.

Moreover, to the best of our knowledge there are no experimental studies in real world open atmosphere conditions to understand the best impact of modulated beam propagation angle fluctuations on the BER of an FSOC which is quantitatively commensurable with the impact of all other factors. Thus, designing a suitable controller to achieve Beam Wandering Compensation (BWC) by processing the beam position information becomes significant. Further, incorporating the developed control system in a real-world open atmospheric FSOC data link and evaluating its greatest contribution by measuring and analyzing the typical communication key parameters are also essential which are the main contributions in this chapter.

6.3 Theory and numerical technique for channel effect and BER evaluation

In the communication point of view, there are many factors that become critical when evaluating the performance of an FSOC system which relies on a traveling wave generated by a laser source at certain transmitted average power level (P_t). The average optical power available at the receiver plane (P_r) at a distance 'R' km in FSOC is estimated by

$$P_r(R) = P_t \frac{D_r^2}{D_t^2 + (R\theta)^2} \exp\left(-\frac{\theta_{pe}^2}{(\theta/2)^2}\right) T_a(R) T_R \,, \tag{6.1}$$

where D_t and D_r are the transmitting optics and receiver telescope aperture diameters respectively, θ is laser beam full-angle divergence, θ_{pe} is the pointing error between the transmitter and receiver Line of Sight (LoS), $T_a(R)$ is the transmittance of the atmosphere along the optical path, T_R is the transmittance of the receiver optics (Das 2012; Afonso et al. 2006). During data transmission, the angle between the transmitter and receiver LoS and the transmitting beam axis must be kept within a fraction of the transmitting beamwidth which may be as small as a few microradians as discussed in Section 6.1. One of the main reasons for the failure of FSOC is absence of beam Pointing, Accusation and Tracking, and (PAT) and therefore, a perfect and continuous adaptive alignment is important to reduce pointing losses (Belmonte & Kahn 2009). In FSOC, the modulated beam cannot be used to measure atmospheric attenuation (Atten) and atmospheric turbulence strength (C_n^2) since it bears PRBS data. One way of simultaneously measuring atmospheric attenuation and turbulence strength is using a separate unmodulated optical link like a transmissometer and scintillometer in parallel to the FSOC data bearing optical beam (Ijaz et al. 2011). This approach is applicable only for indoor test experiments, but in a real-world open atmospheric FSOC system, this approach will increase the overall cost of the system and, furthermore, atmospheric effects mitigation systems are required to stabilize this beam also. Therefore, new models yielding more accurate prediction results according to the local weather data at test fields to measure atmospheric attenuation and turbulence strength are developed and are used to predict these values as given in Eqs. (6.2) and (6.3). The predicted C_n^2 using the new model at the test field is categorized (Kazem 2013) as very weak ($C_n^2 < 10^{-16}$), weak ($10^{-16} \le C_n^2 < 10^{-15}$), moderate ($10^{-15} \le C_n^2 < 10^{-14}$), strong ($10^{-14} \le C_n^2 < 10^{-13}$), and very strong ($C_n^2 > 10^{-13}$).

$$\begin{aligned} \text{Atten} = & -1.09734 - 3.18118\text{Ws} + 2.49896\text{T} + 0.277543\text{RH} \\ & + 0.0667605\text{Ws}^2 + 0.0381112\text{Ws}\,\text{T} + 0.0264432\text{Ws}\,\text{RH} \\ & - 0.106594\text{T}^2 + 0.00641653\text{T}\,\text{RH} - 0.0148677\text{RH}^2 \\ & - 0.0060649\text{Ws}^3 + 0.00118457\text{T}^3 + 9.56075 \cdot 10^{-5}\text{RH}^3 \,. \end{aligned} \tag{6.2}$$

The atmospheric transmittance value is calculated using this equation. The maximum transmittance (T_{max}) is calculated when atmospheric attenuation is minimum, i.e. in very weak atmospheric turbulence condition.

$$
\begin{aligned}
C_n^2 = 1 \cdot 10^{-14}(&5360.63 + 21.0442Ws - 281.763T - 63.5576RH \\
&- 0.0431099Ws^2 - 0.101587Ws\,T - 0.271695Ws\,RH \\
&+ 2.19559T\,RH - 0.26449Ws^3 + 0.199294T^3 + 0.0168798T\,RH^2 \\
&+ 0.000579369RH^3 - 0.001449Ws^4 + 0.0101365Ws^3\,T \\
&+ 0.00092494Ws^3\,RH - 0.00159949Ws^2\,T^2 + 0.000118693Ws^2\,RH^2 \\
&- 0.00265882T^4 - 0.000436822T^3\,RH - 0.000335601T^2\,RH^2 \\
&+ 7.60425 \cdot 10^{-6}RH^3\,Ws - 6.82247 \cdot 10^{-5}RH^3\,T + 1.65979 \cdot 10^{-6}RH^4) \,.
\end{aligned}
\tag{6.3}
$$

Moreover, the magnitude to be considered in designing an FSOC system is the average capacity, which indicates the average best data rate for error-free transmission, the Average Bit Error Rate (ABER) and the outage probability (Ferdinandov et al. 2007). In this work, the Q-factor and BER are theoretically estimated using the calculated transmittance and signal amplitude. These theoretical results are validated with the experimentally measured values. An alternative approach as suggested by Yellow Fourier Technologies (2006) is used to measure the Q-factor and BER performance of the FSOC data link under different environmental conditions by postprocessing the received real-time domain signal and its eye diagrams. The optimum decision threshold (I_{th}) value for data recovery in a NonReturn to Zero–On-Off Keying–Intensity Modulation–Direct Detection (NRZ-OOK-IM-DD) scheme is estimated and adjusted using the mean and variance values of the received signal statistics as in Eq. (6.4) (Freude et al. 2012)

$$
I_{th} = \left(\frac{\mu_0 \sigma_1 + \mu_1 \sigma_0}{\sigma_0 + \sigma_1} \right) \tag{6.4}
$$

where μ_0 and σ_0 are the mean and variance of received bits ‘0’ and μ_1 and σ_1 are the mean and variance of received bits ‘1’.

The performance of an FSOC system is determined by the received Signal to Noise Ratio (SNR) with respect to the total noise at the set (optimum) decision threshold level as in Fig. 6.4. The figure represents the Gaussian distribution of the received signal current for bit ‘0’ and ‘1’ and its corresponding eye diagram of the unipolar NRZ digital data (Freude et al. 2012; Yellow Fourier Technologies 2006; Pedireddi & Srinivasan 2010). Then, the BER is then given by

$$
\text{BER} = P(0)P\left(\frac{0}{1}\right) + P(1)P\left(\frac{1}{0}\right) , \tag{6.5}
$$

where P(0) and P(1) are the probabilities of received bits ‘0’ and ‘1’ respectively. Since the probability of occurrence of receiving bits ‘0’ and ‘1’ is equally probable, $P(0) = P(1) = \frac{1}{2}$. $P(\frac{0}{1})$ is the probability of deciding ‘0’ when actually ‘1’ is transmitted and $P(\frac{1}{0})$ is the probability of deciding ‘1’ when actually ‘0’ is transmitted and given by

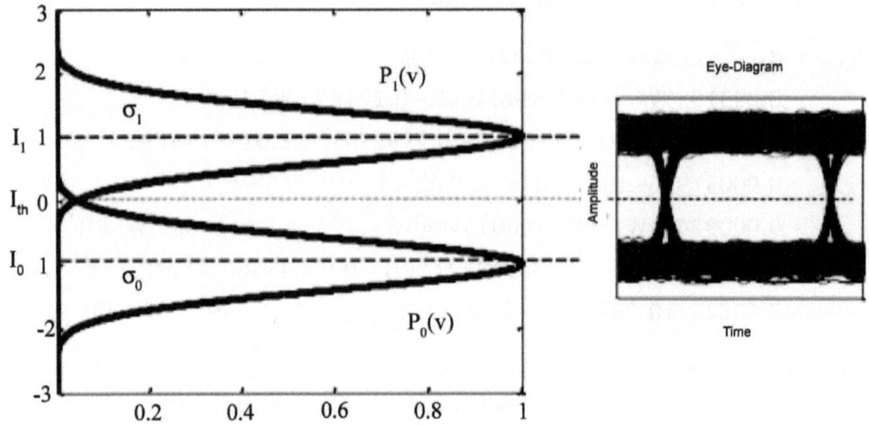

Fig. 6.4: Eye diagram and the distribution of received bits '0' and '1' for BER evaluation.

the Eqs. (6.6) and (6.7) respectively (Mahdieh & Pournoury 2010)

$$P\left(\frac{0}{1}\right) = \int_{-\infty}^{I_{th}} \frac{1}{\sqrt{2\pi\sigma_1^2}} \exp\left(-\frac{(I - I_1)^2}{2\sigma_1^2}\right) dl = \frac{1}{2}\,\mathrm{erfc}\left(\frac{I_1 - I_{th}}{\sigma_1\sqrt{2}}\right) \tag{6.6}$$

$$P\left(\frac{1}{0}\right) = \int_{I_{th}}^{\infty} \frac{1}{\sqrt{2\pi\sigma_0^2}} \exp\left(-\frac{(I - I_0)^2}{2\sigma_0^2}\right) dl = \frac{1}{2}\,\mathrm{erfc}\left(\frac{I_{th} - I_0}{\sigma_0\sqrt{2}}\right) \tag{6.7}$$

In order to correlate the theoretical Q-factor and the resulting BER of the received signal for a range of transmittance values in different turbulence conditions, the maximum Q-factor (Q_{max}) is computed at the receiver as the function of maximum transmittance (T_{max}) and received signal logic '1' and '0' statistics as in Eq. (6.8) (Freude et al. 2012)

$$Q_{max} = T_{max}\left(\frac{I_1 - I_0}{\sigma_0 + \sigma_1}\right) \tag{6.8}$$

where I_1, I_0 are average detected signal current for bit '1' and '0', σ_0, σ_1 are the standard deviations of the noise values for bit '0' and '1'. Then the attenuated Q-factor (Q_{Atten}) can theoretically be approximated with atmospheric attenuation as

$$Q_{Atten} = T_{Atten} Q_{max} \tag{6.9}$$

The unipolar NRZ-OOK-IM-DD modulation scheme is widely used for FSO communication systems because of its ease of implementation, bandwidth efficiency and cost effectiveness (Schneider et al. 2012) and it is used in our work. The theoretical BER for the NRZ-OOK-IM-DD modulation format is given by (Pedireddi & Srinivasan 2010; Freude et al. 2012)

$$\mathrm{BER} = \frac{1}{2}\,\mathrm{erfc}\left(\frac{Q_{Atten}}{\sqrt{2}}\right). \tag{6.10}$$

The Signal to Noise Ratio (SNR) from received signal power and system noise is measured by (Yellow Fourier Technologies 2006; Ali 2014)

$$SNR = \frac{P_{top}^2 - P_{base}^2}{\sum \sigma_{top}^2 + \sum \sigma_{base}^2} \tag{6.11}$$

where P_{top}, P_{base} are the means of most predominant peak of the histogram constructed for high and low logic levels respectively, σ_{top}, σ_{base} are the standard deviations for the means of logic high and low levels respectively. The system metric Q-factor (linear) is measured by (Kumar et al. 2008)

$$Q = \frac{P_{top} - P_{base}}{\sigma_{top} + \sigma_{base}} \tag{6.12}$$

The theoretical BER is estimated using Eq. (6.12) and complementary error function (erfc) as (Kumar et al. 2008; Boffi et al. 2000)

$$BER = \frac{1}{2}\,\text{erfc}\left(\frac{Q}{\sqrt{2}}\right) \tag{6.13}$$

The Eqs. (6.12) and (6.13) are used to measure the data communication link performance in an FSOC system on the basis of signal and noise statistics of received signal and the BER. The link margin (L_M) is therefore calculated by (Bouchet et al. 2006; Freude et al. 2012)

$$L_M = P_t + S_{pd} - L_{tur} - L_{geo} - L_{pe}\,, \tag{6.14}$$

where P_t is transmit optical power, S_{pd} is sensitivity of photodiode, L_{tur} is atmospheric turbulence losses, L_{geo} is optical geometric losses and L_{pe} is pointing error losses.

6.4 Simplex data transmission experimental setup and its description

The FSOC simplex data link experimental setup (transmitter and receiver) as shown in Fig. 6.3 is developed with the necessary optoelectronic components for a link range of 0.5 km at an altitude of 15.25 m. The experimental test bed consists of data source, modulatable Beta-Tx optical source, optical shaker, disturbance sequence generator, Digital to Analog converter (D/A) and transmitting optics at the transmitter station and telescope, Fast Steering Mirror (FSM), variable beam splitter, Optoelectronic Position Detector (OPD), Analog to Digital converter (A/D), Neural-controller (in FPGA) at the receiver. The FSM consists of three terminal piezoelectric actuators based on a moving platform on which the beam steering mirror is mounted. The schematic diagram of the experimental setup is shown in Fig. 6.5. In the experimental setup, all the optoelectronic devices are mounted on vibration damped optical breadboards. A serial Pseudo Random Bit Sequence (PRBS) of $2^{13} - 1$ (Muthiah & Bazil Raj 2012) is given as the input to the optical modulator and the 850 nm, 10 mW optical beam is modulated at the Asynchronous Transfer Mode (ATM) rate of 155 Mbps. Such a high data rate will lead not to use the beacon beam which is normally used to get the beam position and distortion information. The widely used NRZ-OOK-IM-DD scheme, a stream of low (bit '0') and high (bit '1') unipolar NRZ signal are used to directly modulate the optical signal (Hashmi et al. 2010; Zilberman et al. 2010; Singh et al. 2004). The modulating signal amplitude has been set to 100 mV_{pp} in order to achieve a fair comparison.

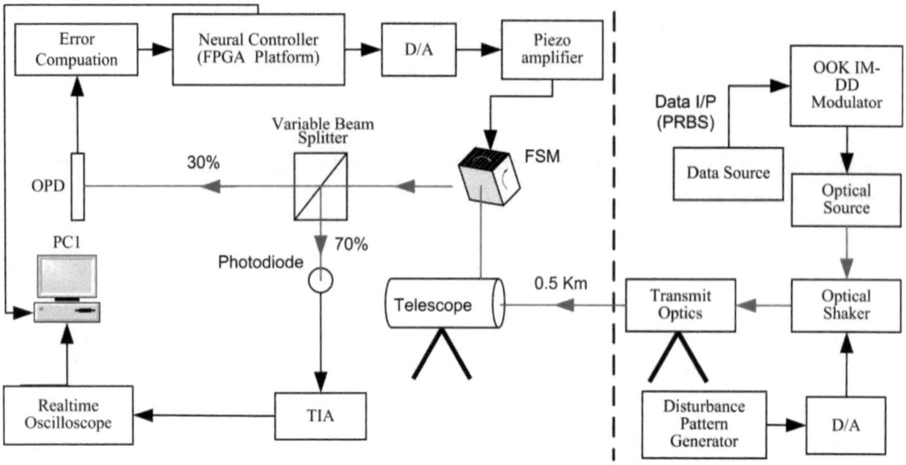

Fig. 6.5: Schematic diagram of the free space optical communication experimental setup for 0.5 km data link. On the left is the receiver with beam steering optoelectronic assembly and ob the right is the transmitter.

The modulated light falls on the optical shaker and is transmitted to the receiver through the atmospheric turbulent optical channel. Transmitting optics are used to increase the beam diameter from 3 mm to 9 mm. An arbitrary disturbance sequence generator circuit is developed in the FPGA and corresponding analog voltages are given to the optical shaker on which a Pure Reflection Mirror (PRM) is mounted to act as the second possible source of jitter. The intensity modulated optical beam propagates through the real-world open atmosphere and is detected at the optical receiver. The telescope captures all the optical power and reflects it to fall on the FSM. The incident laser beam of the FSM gets reflected to the variable beam splitter that splits the incident into two beams: reflecting beam and propagating beam. The reflected beam is made to fall on the optical detector (photodiode) followed by a Trans Impedance Amplifier (TIA) used for communication purposes. The equivalent photocurrent at the output of the photodiode is amplified using a TIA. The output of the TIA is captured using a high-frequency digital oscilloscope where full signal analysis (i.e. the eye diagram, received signal histograms, the Q-factor, and BER) is carried out in different turbulence conditions, i.e. from very weak to very strong atmospheric turbulence strength. A first order low pass filter with a cut-off frequency of $0.75 \cdot |R|$ Hz is used to reduce noise, where $R = 1/T_b$ and T_b is the minimum pulse duration (bit time) (Zhao et al. 2010a). The propagating beam (passed through the beam splitter) falls on the OPD. The outputs of the OPD are applied to the signal conditioning and error computation unit and the error data is given to the neural-controller developed in FPGA. The controller output data is converted to an analog signal using the D/A converter and given to the FSM through the piezo-amplifier. The measurement data, control process performance and the quality metrics of data link are continuously monitored using PC1. The OPD outputs are normalized to the maximum intensity, i.e. sum of energy of all the quadrants. Therefore, the primary candidate of interest in this work is not the scintillation index but only the beam centroid information. Further, the shape (4 mm diameter) of the received beam falling on the OPD plane is kept unchanged using the optical collimator at the eye-piece of the telescope.

6.5 Experimental results and data analysis

The diurnal period is classified as three sessions: (1) morning (6.00 to 12.00 Hrs), (2) afternoon (12.00 to 18.00 Hrs), and (3) night (18.00 to 6.00 Hrs). The BER is continuously measured in all the sessions and the results are recorded in the data logging file. The session-wise minimum, maximum and mean BER corresponding to the experimental data measured in three different days are given in Tab. 6.2. This measurement is carried out at the data rate of 155 Mbps and in the absence of the BWC control. The measurement data clearly evidences that the measured BER mostly fluctuates continuously, which degrades the overall performance of the FSOC data link and decreases the reliability and quality of the communication system. This happens purely because of the received power fluctuation at the detector plane due to scintillation and beam wandering as shown in Fig. 6.5. Therefore, an attempt has been made to change this trend using a BWC control (neural-controller as observer based closed loop feedback controller) system in different local weather conditions. The outstanding performance of the neural-controller in improving the received signal quality is analyzed in a communication perspective.

The daylong behavior of atmospheric turbulent channel effects on the modulated optical signal is both theoretically estimated and practically measured simultaneously. These results are used to comprehensively estimate and analyze the degree of correlation between them to understand the contribution of the developed BWC control system and BER calculation method. The significance of the developed controlling scheme in improving the overall stability and reliability of the FSOC data link is evaluated and the results are presented in this section. The communication metrics such as received signal power, transmittance, eye diagram, Q-factor, BER and link margin values are continuously measured with and without BWC control in different

Tab. 6.2: BER summary of FSOC data transmission experiments and results.

Date (Session)	Min BER	Max BER	Mean BER	Comments on BER trend
17/12/13 (1)	$1.80 \cdot 10^{-7}$	$4.19 \cdot 10^{-7}$	$2.06 \cdot 10^{-7}$	Stable, slightly decreasing
17/12/13 (2)	$9.64 \cdot 10^{-7}$	$3.72 \cdot 10^{-6}$	$8.06 \cdot 10^{-7}$	Variable ↑↓
17/12/13 (3)	$1.67 \cdot 10^{-7}$	$4.38 \cdot 10^{-5}$	$2.18 \cdot 10^{-6}$	Variable monotonically decreasing
23/01/14 (1)	$5.74 \cdot 10^{-7}$	$4.93 \cdot 10^{-4}$	$8.43 \cdot 10^{-5}$	Variable, conspicuous bump
23/01/14 (2)	$1.51 \cdot 10^{-6}$	$1.54 \cdot 10^{-6}$	$0.38 \cdot 10^{-6}$	Stable with a final increase
23/01/14 (3)	$4.35 \cdot 10^{-5}$	$4.13 \cdot 10^{-3}$	$2.16 \cdot 10^{-4}$	Stable with a final exponential increase
06/02/14 (1)	$3.64 \cdot 10^{-5}$	$3.63 \cdot 10^{-5}$	$3.61 \cdot 10^{-5}$	Only two measurements
06/02/14 (2)	$6.45 \cdot 10^{-9}$	$6.45 \cdot 10^{-9}$	$6.40 \cdot 10^{-9}$	Perfect link: BER ≈ 0
06/02/14 (3)	$5.78 \cdot 10^{-5}$	$4.81 \cdot 10^{-8}$	$6.21 \cdot 10^{-6}$	Variable ↑↓

turbulent conditions and some of the important results are reported and analyzed through which the potential and feasibility of the proposed neural-controller are highlighted.

6.5.1 Comparative evaluation of received signal statistics

The atmospheric transmittance values are calculated from the attenuation values estimated using Eq. (6.2) and simultaneously the ensemble average values of the received signal are recorded. The modulation voltage of 100 mV_{pp} is used for compression convenience and the average transmitted optical power is maintained unchanged during the experimentation. This implies that the amplitude level and the time-slot duration of OOK-NRZ modulation data are constant at the transmitter station. The normalized received power against a range of transmittance is shown in Fig. 6.6. According to this figure, the average power level decreases not only due to the scintillation effects but also due to beam centroid wandering, especially in the low transmittance regime. Further, the BWC control becomes insignificant at the maximum transmittance ($T_{max} \approx 90\,\%$) and its normalized power level is 0.84 in both cases. The power level dramatically decreases with increasing atmospheric attenuation, i.e. decreasing the transmittance when the BWC control is turned off. The power levels are significantly improved by using the BWC control system as in Fig. 6.6, which exhibits the importance of the beam stabilizing system. In the presence of BWC control, the normalized power level decrement is considerably deviated for the transmittance value of ≤ 86 %. For example, in the absence of BWC control, the normalized power level is ≈ 0 at transmittance of ≈ 10 % while it was 0.213 for the similar transmittance value in the presence of BWC control. These results confirm that the power losses due to pointing error, as in Eqs. (6.1) and (6.14), is almost nullified.

The eye diagram is constructed as suggested by Yellow Fourier Technologies (2006) by postprocessing the received signal obtained from the output of the low pass

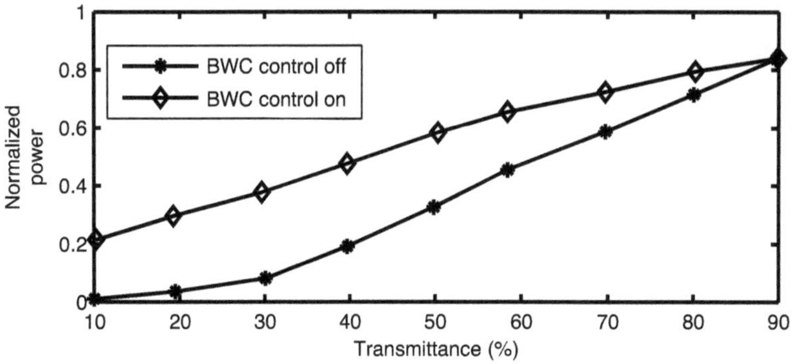

Fig. 6.6: Normalized received power against a range of transmittance with and without Beam Wandering Compensation (BWC) control.

filter (0.75 · |R| Hz) which corresponds to the transmitted data at the transmission speed of 155 Mbps as shown in Fig. 6.7 (a) and (b). The bit time (T_b) is 6.45 ns with peak-to-peak amplitude of 0.64 V and all measurements are carried out with respect to these values. The width of the eye-opening is measured in different atmospheric turbulence conditions and a comparatively wider eye-opening is observed in the presence of BWC control even in strong atmospheric turbulence condition as in Fig. 6.7 (a) and (b) for the identical noise variance. The beam wandering of the received optical signal results in decreasing the width of the eye-opening and average amplitude considerably. This in turn causes the distribution of bits '1' and '0' to be more flat due to the loss in the eye-opening (height) of the eye diagram. For example, in Fig. 6.7 (a) the peak-to-peak amplitude is ≈ 0.21 V whereas in Fig. 6.7 (b), it is ≈ 0.40 V. Further, the timing response of the constructed eye diagram exhibits that the amplitude fluctuates more, once beam wandering is unmitigated. This is due to the substantial increase of eye width at the bits '1' and bits '0' of the eye diagram. The link availability is achieved with an OOK-NRZ modulation scheme at the BER variations in between $6.45 \cdot 10^{-9}$ and $7.09 \cdot 10^{-8}$ in the strong atmospheric fluctuation in the presence of BWC control. Note that the eye height is decreased to 0.013 V without BWC control and noise becomes the dominating factor, while it is 0.36 V in the presence of BWC control with negligible noise contribution which results in a considerable level of improvements on signal stability at the detector plane. These statistics indicate that the OOK-NRZ signaling format for the FSOC data link is more robust when beam wandering is mitigated. All the necessary parameters required to estimate the Q-factor and BER are measured from the constructed eye diagrams.

The Q-factor variations of the received signal for the FSOC system with OOK-NRZ modulation format is investigated as a function of transmittance in the presence of BWC control. The Q-factor measurement has been extensively used to characterize the quality of the optical communication system especially in fiber optics as reported in

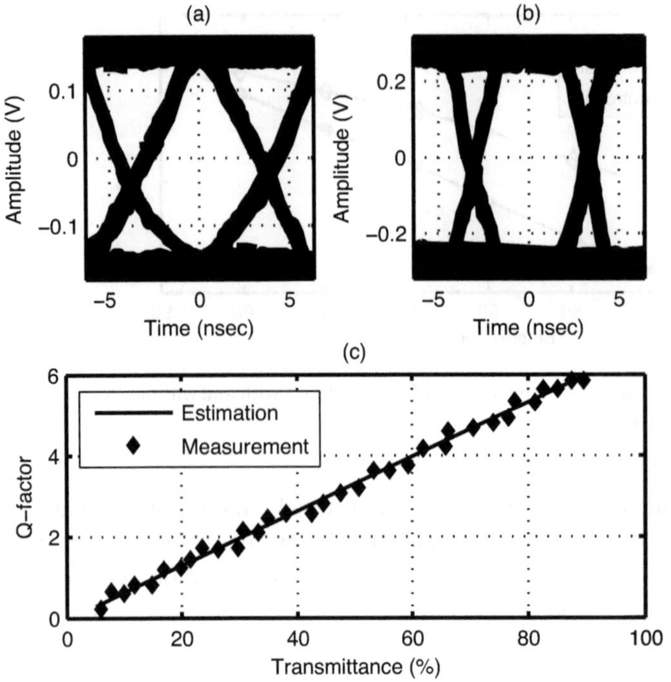

Fig. 6.7: Postprocessed eye diagrams for same modulation scheme with BWC control (a) off and (b) on conditions. (c) Q-factor versus transmittance for OOK-NRZ scheme: theory (solid lines) and experiment (dots).

Ohteru and Takachio (1999), Matera and Settembre (2000) and Shake et al. (2001). The received signal amplitude corresponding to the transmitted data from very weak to very strong atmospheric turbulence conditions are measured and recorded. The experimental Q-factor values obtained using Eq. (6.12) have been recorded, comparatively analyzed with the values of Q-factor estimated using Eq. (6.9) and the results are shown in Fig. 6.7 (c). The experimental results exhibit a good agreement with the estimated values and show that the OOK-NRZ signaling format offers the best Q-factor against a range of values of transmittance due to its high Peak to Average Power Ratio (PAPR). Furthermore, the results showed that the Q-factor decreases linearly with increasing atmospheric attenuation for the NRZ-OOK modulated beam due to atmospheric absorption and scattering. Once beam wandering is unmitigated, the Q-factor varies steeply with the mean value of 1.93 due to the wander of the Gaussian beam centroid in addition to the scintillation effects.

The performance of BER against the range of transmittance (0 % to 90 %) for the OOK-NRZ scheme is shown in Fig. 6.8 and the BER axis is truncated to $9.36 \cdot 10^{-9}$. The measured (using Eq. (6.13)) BER values keep a good agreement with the theoretically estimated BER values. The theoretical curve is calculated using Eq. (6.10).

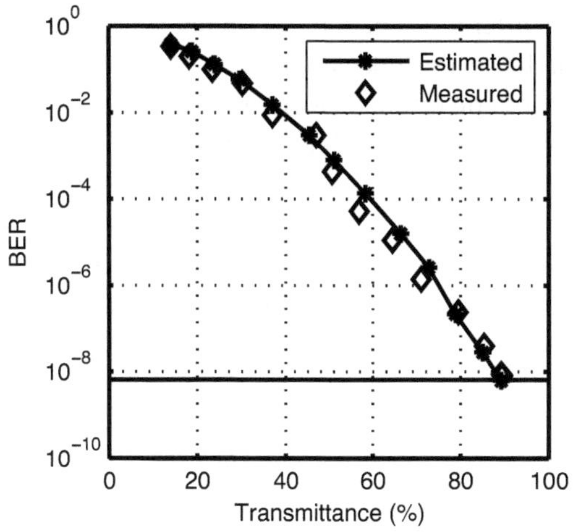

Fig. 6.8: BER against a range of transmittance of an FSOC channel.

This measurement is carried out with constant input average power of 10 mW from the transmitter station and $Q_{max} \approx 6$ for T_{max} of 90 %. When beam wandering is unmitigated, the beam centroid starts fluctuating, so that transmittance decreases. It is observed that the link availability at a BER value of 10^{-6} is related to the Q-factor of ≈ 4.82 following Eq. (6.13). The required value of Q-factor to maintain the maximum availability i.e. 99.999 % cannot be achieved under all weather conditions without a beam steering system, even with increased input power (Mahdieh & Pournoury 2010; Liu et al. 2014).

The link margin (L_M) is estimated according to Eq. (6.14) using average transmit optical power (P_t) of 10 mW, photodetector sensitivity (S_{pd}) of 0.59 A/W at 155 Mbps, fluctuating atmospheric turbulence loss (L_{tur}), fixed (constant) geometrical loss (L_{geo}) of 0.0023 mW and pointing error (L_{pe}) and the results are shown in Fig. 6.9. The losses due to atmospheric turbulence are obtained by estimating atmospheric attenuations using Eq. (6.2) in all weather conditions. This computation is carried out with the pres-

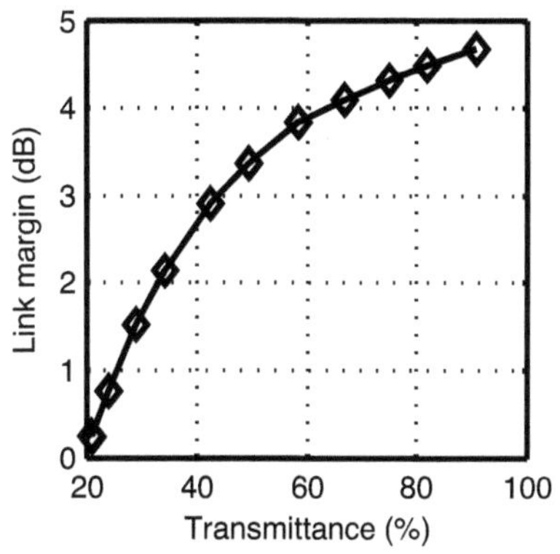

Fig. 6.9: Link margin values against the transmittance estimated under different atmospheric turbulence strength conditions with BWC control.

ence of BWC control, therefore the fourth term in Eq. (6.14) becomes negligible i.e. almost zero during the experimentation. Hence, the link margin computation becomes a function of only one variable i.e. atmospheric turbulence losses. The experimental results obtained in different weather conditions suggest that a reliable FSOC link can be established for the link range of 7 km with the BWC control system at the data rate of 155 Mbps in a range of atmospheric turbulence conditions varying within very weak and strong regimes. However, in very strong atmospheric turbulence conditions, the link margin becomes insufficient (outside the margin) to maintain a reliable and error free data transmission which leads to wavefront local tip/tilt corrections towards the further improvements.

6.5.2 Impact validation of beam wandering compensation system

As atmospheric turbulence strength increases, the propagating modulated optical beam will experience larger degradation, centroid wandering and break-up from the transmission axis which will disconnect the data reception and its controllability. The profile of the received beam centroid wander is recorded using the science camera in IR mode at the output (eye-piece) of the telescope to understand beam wandering displacement. The classifications of turbulence strength estimated using the developed model as in Eq. (6.3) on 17th January 2014 (Friday) is moderate and strong. The Gaussian beam of diameter of 4 mm beam profile recorded on the same day with the BWC on and off conditions is shown in Fig. 6.10 (a) and (b) respectively. The beam still retains its characteristics, shape and is well centered once the BWC control is turned on, otherwise the beam experiences degradation and loses its position stability as shown in Fig. 6.10 (b). The centroid coordinates are off center from −0.2 to 1.2 mm respectively in the x and y directions due to this specific realization. In stronger turbulence, the beam experiences a strong degradation and starts breaking up the link when beam wandering is unmitigated.

Geometrical interpretation of the wavefront (global) of the received optical beam at the communication detector plane is modeled using the beam centroid position

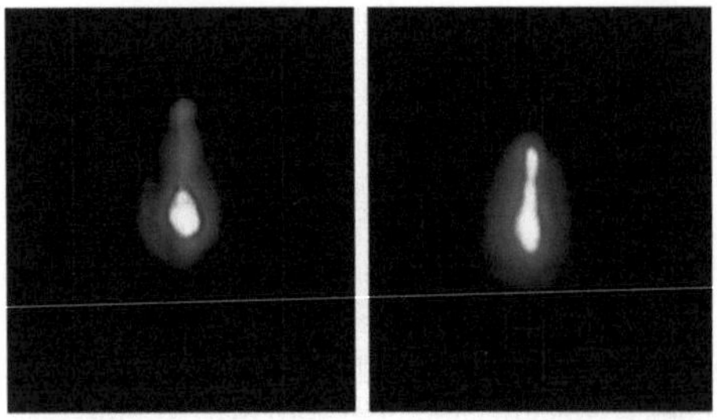

(a) (b)

Fig. 6.10: Illustrations of beam profile when beam wandering (a) mitigated and (b) unmitigated after propagating 0.5 km. The axes are compressed mm scale.

information: x, y, and radial distance, obtained from the OPD using the geometrical theory suggested in Zheng (2014), Zocchi (2005) and Leitgeb et al. (2009). The time series geometrical interpretations of beam wavefront (global) tilts is estimated on 5th February 2014 (Wednesday) and are shown in Fig. 6.11. In Fig. 6.11, (1, 1) illustrates the beam wandering mitigated wavefront profile while the others correspond to the unmitigated wavefront and it clearly exhibits that the beam centroid randomly wanders on the detector plane. These results evidence that the beam wandering compensation system becomes significant to couple the Power In the Bucket (PIB), i.e. telescope collected power, to the communication detector for the successful installation of a reliable FSOC system.

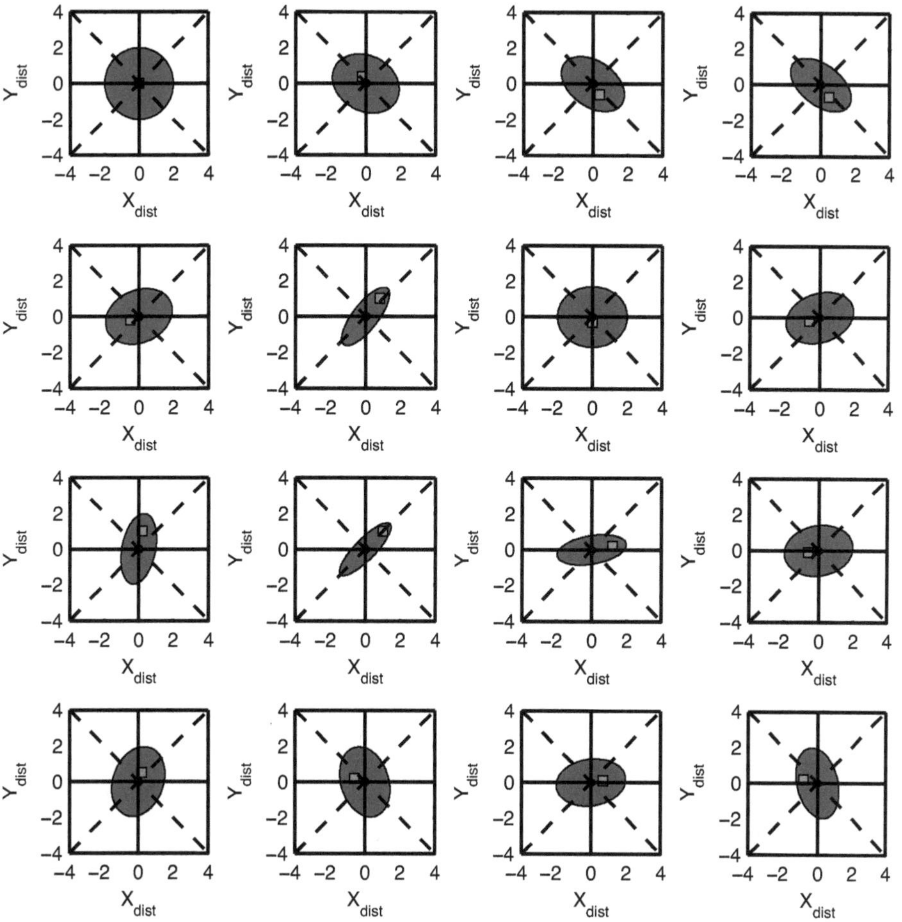

Fig. 6.11: Geometrical interpretation of beam wavefront (global) profile on the detector surface after propagating 0.5 km horizontal optical path under different turbulence conditions. Beam spot and radial distance are illustrated by red and green colors respectively.

Separate experimental studies to quantitatively measure the dependency of the Q-factor and BER in terms of beam centroid displacement in different local atmospheric turbulent conditions were conducted to understand the overall behavior of the FSOC channel at the installation field. For accomplishing this experimental study, the radial displacement distance of the beam centroid on OPD plane is divided into six regions and denoted by the linguistic terms: VS (0–0.5 mm), S (0.5–1 mm), M (1–1.5 mm), H (1.5–2 mm), VH (2–2.5 mm), and VVH (2.5–3 mm), where VS is Very Small, S is Small, M is Medium, H is High, VH is Very High and VVH is Very Very High. The Q-factor is measured as given in Eq. (6.12) for different regions (VS through VVH) of the radial displacement on the OPD plane, the corresponding theoretical BER is estimated using Eq. (6.13) and the results are shown in Fig. 6.12. The received power fluctuates due to the scintillation effects also in addition to beam wandering. The results show that the Q-factor value linearly decreases and the BER increases with increasing centroid radial displacement distance of the received beam.

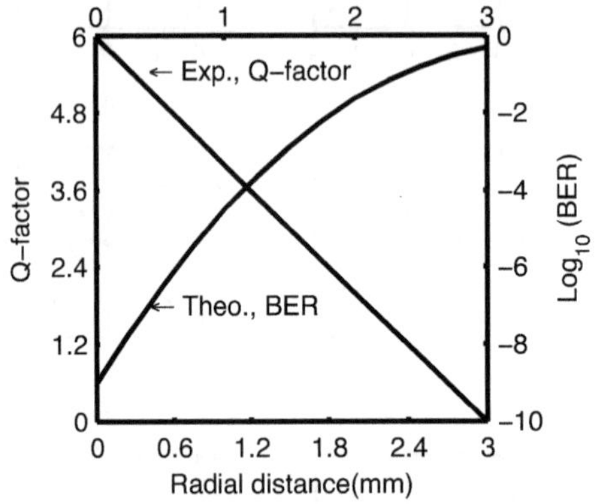

Fig. 6.12: Experimental Q-factor and theoretical BER estimation against beam centroid displacement on OPD.

This is due to the outer scale variations on the intensity of the optical signal falling on the communication detector. These results exhibit the importance of beam stabilization in the FSOC system and mean that the received beam centroid must be stabilized regardless of its Angle of Arrival (AoA). The radial displacement distance values measured on 2nd December 2013 (Monday) for the diurnal period are used to generate the histogram plot to identify the most occurring sample values and their corresponding linguistic terms (VS through VVH). The most frequent radial displacement distances and their linguistic terms with minimum and maximum values of Average BER (ABER) are given in Tab. 6.3. Almost the same results were reported in different weather conditions.

The daylong measurements clearly demonstrate that the received beam centroid must be controlled within the VS region. The improvements on the stability of the Q-factor and corresponding BER due to the beam steering control system are experimentally measured and the results are shown in Fig. 6.13. Q-factor and BER measure-

Tab. 6.3: Experimentally measured radial displacement and BER statistics.

High frequency samples (mm)	Radial distance linguistic	ABER	
		Min	Max
0.143	VS	$6.45 \cdot 10^{-9}$	$1.04 \cdot 10^{-8}$
0.614	S	$7.13 \cdot 10^{-7}$	$7.81 \cdot 10^{-5}$
1.235	M	$8.51 \cdot 10^{-5}$	$5.31 \cdot 10^{-3}$
1.846	H	$5.83 \cdot 10^{-3}$	$1.41 \cdot 10^{-2}$
2.314	VH	$2.45 \cdot 10^{-2}$	$6.25 \cdot 10^{-1}$
2.828	VVH	$6.79 \cdot 10^{-2}$	$1.34 \cdot 10^{-1}$

ments are performed with the beam steering system alternately turned-off and -on environment in the one minute time interval. Theoretical BER is also estimated to understand the correlation with the practically measured values. The results exhibit that the measured BER keeps close agreement with the theoretical BER. The BER in Fig. 6.13 is truncated to $6.46 \cdot 10^{-9}$ i.e. one bit error.

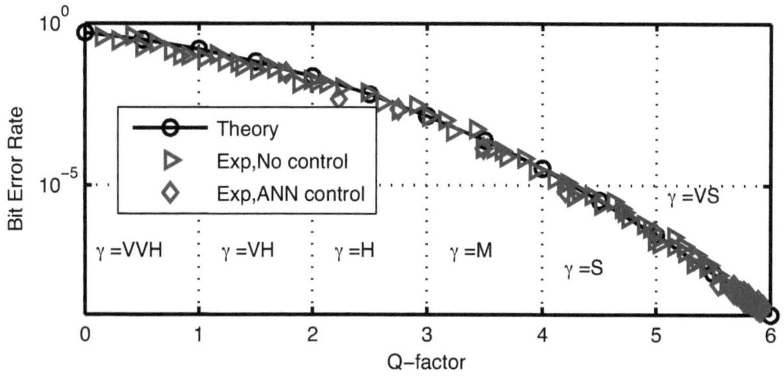

Fig. 6.13: BER as a function of Q-factor. Theoretical (black) and measured BER with beam steering feedback turned off and on alternately for every 60 s time interval.

The variation range of Q-factor in no control is from 0 to 5.6 and BER is from $9.32 \cdot 10^{-1}$ to $1.05 \cdot 10^{-8}$, whereas in control Q-factor is from 5.4 to 6 (maximum value) and BER is from $6.45 \cdot 10^{-9}$ to $1 \cdot 10^{-8}$ which is less than 10^{-6}. Once the control is on, beam wandering is controlled within the VS region so that the Q-factor value is improved and hence BER is decreased below 10^{-8}, otherwise, the beam wanders throughout all the regions S through VVH and BER falls in the upper regime. Furthermore, BER reaches greater than 10^{-6} even with beam steering control at the beginning of measurement (for few runs) due to bit slip and false detection during the pilot sequence synchronization scanning time in the real-time oscilloscope. As can be seen from Fig. 6.13, the greatest improvement in communication quality compared to the no control experiment is demonstrated, beam stability is improved, the outage probability is decreased and better BER is achieved without any Forward Error Control (FEC) techniques.

6.5.3 Quantitative analysis of atmospheric turbulence effects on communication parameters – improvement and reliability

In order to quantify the FSOC system performance with beam wandering compensation (BWC) control in different atmospheric turbulence strength (C_n^2), the Q-factor is measured using the postprocessed eye diagram using Eq. (6.12). The measured Q-factor against a range of usually occurring C_n^2 values at the test field is shown in Fig. 6.14 with beam wandering compensation on and off conditions. For the clarity of the figure, the x-axis is taken in the logarithmic scale and the initial value of C_n^2 is selected to be 10^{-16} rather than zero for obtaining Q_{max} in very weak atmospheric turbulence conditions. In both case, the measured Q_{max} is 6 and the values of the Q-factor decrease with increasing C_n^2. The Q-factor decreases rapidly when beam wandering is unmitigated. The trend is utterly changed and the greatest improvements are obtained in Q-factor decrement in the presence of BWC control. For example, when the $C_n^2 = 1 \cdot 10^{-14}\,m^{-2/3}$, the Q-factor values with and without BWC control are 4.3 and 1.8 respectively. Further, the Q-factor reaches 0 when the value of C_n^2 is $\approx 0.5 \cdot 10^{-13}\,m^{-2/3}$ in the absence of BWC control and the C_n^2 value is $\approx 6 \cdot 10^{-13}$ for the Q-factor value of 0 in the presence of BWC control. These results evidence that, due to the atmospheric turbulence strength fluctuations, scintillation is not the only degrading parameter but also beam wandering. Further, compensating beam wandering in different atmospheric conditions significantly increases the received signal power level and stability at the detector plane.

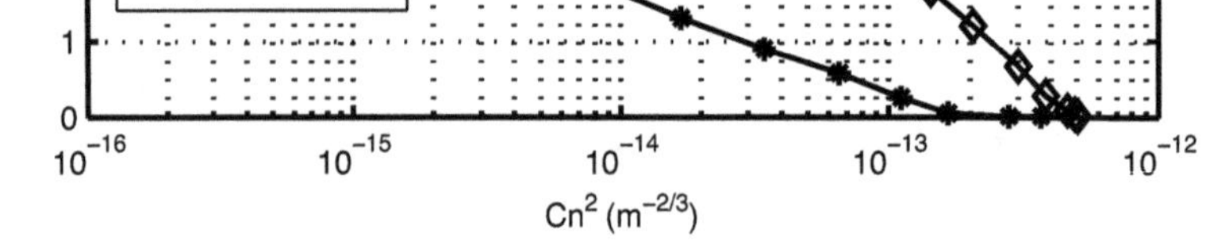

Fig. 6.14: Experimental Q-factor variations against a range of C_n^2 with beam steering system off and on conditions.

The analysis on the impact of turbulence strength fluctuations on the fixed decision threshold value estimated using Eq. (6.4) is carried out by recording the distribution of the received signal obtained from the output of the TIA. This can be described by marking 0 V as the decision threshold level as in Fig. 6.15 (a) and (b). The received signal distribution for bits '1' and '0' are equally spaced and comparable for both sides of the decision threshold level mark and the bits '1'and '0' are clearly distinguishable in the presence of the BWC control even in the strong turbulence condition as in Fig. 6.15 (b). In the absence of BWC control, the received signal distribution becomes distorted as in Fig. 6.15 (a), even in weak turbulence conditions. The received signal distribution is heavily distorted and no longer distinguishable in strong and very strong atmospheric turbulence conditions when beam wandering is unmitigated, which results in a decreased Q-factor and increased BER values. As in Fig. 6.15, the maximum count values of ≈ 290 out of 2000 ensembles are observed at the voltage level of −0.28 V for bit '0' and 0.25 V for bit '1' with many voltage level distributions in between these levels when BWC control is off; whereas in on, the observed maximum count values are ≈ 590 for bit '0' at −0.3 V and ≈ 490 for bit '1' at 0.3 V out of 2000 ensembles with the ≈ 0 V level distribution in between these levels.

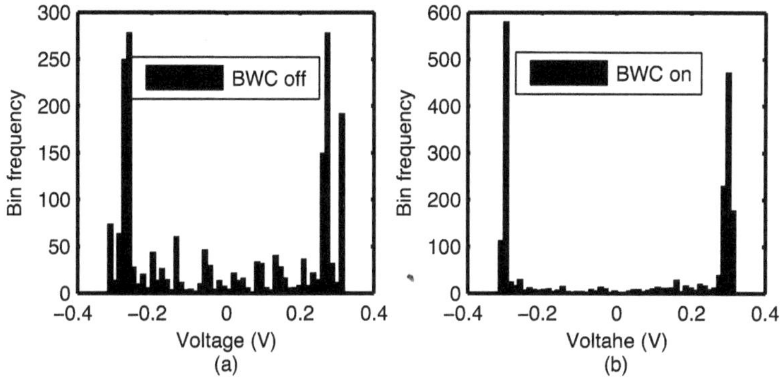

Fig. 6.15: Histograms of OOK-NRZ received signal with beam wandering compensation control (a) off in weak and (b) on in strong atmospheric turbulence conditions.

Typical values of BER versus different atmospheric turbulence conditions characterized by C_n^2 for 850 nm and 155 Mbps data link with BWC control on and off conditions are shown in Fig. 6.16 and its y-axis scale is truncated to the minimum possible value of $6.45 \cdot 10^{-9}$. From this figure, in general, it can be seen that BER has very small values at the turbulence with weak refractive index fluctuation and rises dramatically with the C_n^2. According to this result, BER becomes very large at turbulence with very strong refractive index fluctuation, i.e. $C_n^2 > 6 \cdot 10^{13}\,\mathrm{m}^{-2/3}$. Further, the results show that BWC control is very important for possibly reducing the BER to some further extent. As the figure illustrates, for similar atmospheric turbulence conditions, the incorporation of BWC control significantly reduces BER. For example, in the absence of BWC control, BER reaches the minimum value i.e. $6.45 \cdot 10^{-9}$ only when the C_n^2 value lies in between $8 \cdot 10^{-16}\,m^{-2/3}$ and $3 \cdot 10^{-15}\,m^{-2/3}$ whereas in BWC control this happens for the C_n^2 value varying in between $1 \cdot 10^{-14}\,m^{-2/3}$ and $5 \cdot 10^{-14}\,m^{-2/3}$. The BER in a similar turbulence interval falls in the range of $1 \cdot 10^{-6}$ and $1 \cdot 10^{-1}$ in the absence of BWC control. The greater improvements can be noticed in the figure starting from strong turbulence condition onwards. These results show that the FSOC performance and reliability are improved significantly. However, at turbulence with very strong refractive index fluctuation, the influence of BWC control on BER becomes insignificant and the difference becomes smaller. It must be noted that in commercial FSOC systems a typical value of BER $\lessapprox 10^{-6}$ is acceptable for reasonably good performance. Thus, BWC control works perfectly even in stronger turbulence conditions unlike the case without BWC control.

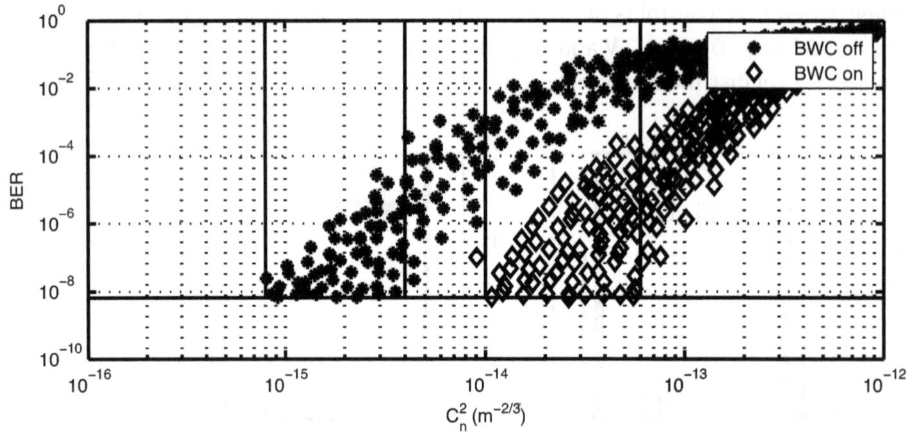

Fig. 6.16: BER versus estimated a range of C_n^2 with beam wandering compensation control on and off conditions.

The influence of BWC control in a daylong operation is investigated by subsequently turning on and off the feedback control in the interval of 60 s, as suggested by Yoon-Suk Hurh et al. (2005), and simultaneously the BER values are recorded. A portion of the experimental results for 600 runs in the total measurement time of 10 minutes is shown in Fig. 6.17. Note that the zero BER is truncated to the smallest possible nonzero value of BER i.e. $\approx 6.45 \cdot 10^{-9}$ which corresponds to a single error bit in a 155 Mbps data transmission system. As can be seen from Fig. 6.17, the measured BER in the given interval without BWC control reaches a maximum of four orders of magnitude higher than that with BWC control. BER varies in the inner scale and is controlled within $\approx 6.45 \cdot 10^{-9}$ and $\approx 7.09 \cdot 10^{-8}$ with the mean value of $3.86 \cdot 10^{-8}$ which is less than

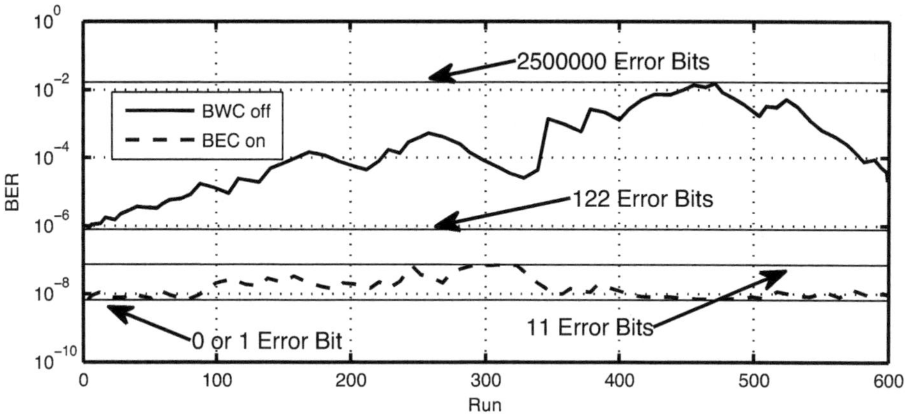

Fig. 6.17: Measured BER in 600 runs with alternately beam wandering compensation turned on and off conditions with the measurement interval of 60 s. The horizontal lines indicate the lower and upper boundary of measured BER in both cases.

10^{-6}. In the absence of BWC control, BER varies over the outer scale and measured min/max values are $7.89 \cdot 10^{-7}$ and $1.6 \cdot 10^{-2}$ with the mean value of $8 \cdot 10^{-3}$. These results exhibit that the quality and reliability of a terrestrial FSOC data link at 155 Mbps is improved in a great manner in an atmospheric turbulence strength from very weak to strong. The availability of an FSOC link maintaining a BER of 10^{-6} in different atmospheric turbulence strengthes is possible only with the BWC control system.

However, the performance of the data link deteriorates drastically as the atmospheric turbulence strength increases to a very strong condition and BER increases to $\approx 10^{-2}$. Sometime we have observed that the link availability becomes a problem and connectivity fail in the very strong turbulence regime. Incorporating high speed signal conditioning (Texas Instruments 2013), wavefront local tip-tilt sensing and high speed wave front aberration (Phase distortion) correction techniques (Planchon 2002) will be finished in the near future. A new version of the neural-controller design for the deformable mirror control (≈ 32 piezo actuators) and its implementation in FPGA is the matter of subsequent research.

6.6 Summary

Various advantages and features of the FSOC system are described. The influence of turbulent atmospheric parameters on the propagation of the modulated optical wave is explained. The state of art and the literature survey results are reviewed. Atmospheric attenuation and turbulence strength are estimated using the developed new models. The Q-factor and BER are estimated from the statistics of eye diagram constructed by postprocessing the received signal. The construction of the FSOC transmitter and receiver experimental setups at transmission speed of 155 Mbps for the link range of 0.5 km at an altitude of 15.25 m are described with the principle of operations of all the optoelectronics devices. The BER profile is recorded for the diurnal period without the BWC control system and the outer scale variations from $6.45 \cdot 10^{-9}$ to $3.45 \cdot 10^{-3}$ are observed. The received signal ensemble average power is measured for a range of transmittance with and without BWC control and the greatest improvement is observed in the presence of BWC control. The power fluctuations due to beam centroid wandering is mitigated, so that the average received power is improved through which the transmittance and Q-factor are significantly increased. In almost all weather conditions, transmittance is maintained within 48 % and 90 %, so that the Q-factor is obtained in between 2.9 and 6. The eye-opening height is almost maintained at $\approx 0.6\,V_{pp}$ in almost all weather conditions in the presence of BWC control whereas in its absence the value varies between $0.013\,V_{pp}$ to $0.56\,V_{pp}$, which leads to improper decisions while decoding the data.

BER performance is theoretically and experimentally evaluated for the OOK-NRZ format at the ATM line data rates for different transmittance values and the correlations are analyzed. The link margin is calculated as a function of atmospheric turbulence effects with nullifying the pointing error loss and found that with the same specifications, data link at 155 Mbps could be achieved for the maximum range of 7 km. Beam centroid wandering is studied with a science camera in IR mode and a long-term beam wandering profile is recorded. A geometrical interpretation of the beam wavefront on the detector plane is modeled and predicted as a function of beam position information. The dependency of the Q-factor and BER with the beam centroid radial displacement is intensively investigated and found that the BER of $6.45 \cdot 10^{-9}$ can be maintained by positioning the beam centroid within the VS region. Similarly, the dependency of the Q-factor and BER with the C_n^2 are also studied and their significant improvements are analyzed. The histogram plot corresponding to the received data signal is generated and the distributions of the received bits '0' and '1' are analyzed. The results exhibited that the distributions are greatly distinguishable when beam wandering is mitigated, otherwise distinguishing the logic levels and arriving at decision on decoding becomes difficult, which leads to increased BER and misdetection. The results presented in this chapter are intended for setting up the FSOC transceiver.

7 Conclusions and future work

7.1 Conclusions

In recent years, the demand for a back-up and complementary link to the Radio Frequency (RF) technology mainly for the last-mile access in networks based on the FSOC system has increased significantly. This is due to a number of advantages and/or features including a large unregulated and license-free transmission bandwidth spectrum, large data transmission which consumes low power, security as well as immunity to electromagnetic interference. However, the constraints imposed by the atmospheric channel such as turbulence and environmental changes decrease the system's performance and availability of the link. The modeling of atmospheric attenuation, turbulence strength and characterization of the FSO channel are important and need to be investigated to enhance the robustness of FSO links.

This thesis commenced from the brief overview on FSO communications in Chapter 1. Background, research motivation, applications and advantages/features of FSO technology make it more viable compared with the existing RF technologies and its potential areas of application were discussed. The basic concept of FSO communication with a conceptual diagram has been explained. The main characteristics: directionality of the light beam, form factor, wavelength selection criteria, challenges and limitations and different effects such as absorption, scattering and atmospheric turbulence posed by the atmospheric channel on the transmitted optical beam have been introduced. Also, the research objectives, original contribution/newness and achievements are discussed in Chapter 1.

The importance of meteorological measurements for FSOC channel behavior modeling is introduced with the literature review results of background and related works. The construction of the field test experimental setup is explained. The meteorological parameters measurement protocols: digital counting, two wire interface and serial peripheral interface are introduced. The digital architectures developed in FPGA for interfacing the specialized weather sensors to measure the meteorological parameters are explained with the finite state machine, pseudocodes and simulation diagram in Chapter 2. Data logging communication protocol is discussed within the communication frame format. The performance calibrations of the developed measurement system and uncertainty computations are detailed. The highest correlation coefficients of $R = 0.9992$ and $R = 0.9988$ are achieved for wind speed and pressure measurements and the correlation coefficients of $R = 0.9963$ and $R = 0.9973$ are obtained for temperature and relative humidity measurements respectively. The mathematical background related to the PAMELA model for predicting turbulence strength is reviewed. Some of the experimental results exhibiting reasonable correlation with the measurement data for different seasons in a one year period are presented in Chapter 2.

The essentials of modeling atmospheric attenuation and turbulence strength are introduced with the practical measurements of visibility and optical attenuation. The literature review results are discussed. The optoelectronics assembly comprising transmitter and receiver built and used for suitable data acquisition for modeling atmospheric attenuation and turbulence strength is explained function-wise by using a block diagram. Different models existing for atmospheric attenuation and turbulence strength prediction are experimentally tested and based on the prediction results and measurement values, some of the models exhibiting less prediction error have been selected for the comprehensive analysis. These selected models are reviewed with their limitations and detailed in Chapter 3. The formulation of separate mathematical models for atmospheric attenuation and turbulence strength based on the response surface of the experimental data acquired from the real-world open atmosphere at the test-field and linear regression analysis are explained. The developed atmospheric attenuation and turbulence strength models prediction accuracy are investigated using the ANOVA tools and based on the ANOVA investigation results only two models (each one): cubic equation ($R^2 = 98.76\,\%$) and model equation V ($R^2 = 98.93\,\%$) have been conformed. The confirmatory test has been conducted to ensure the prediction accuracy of the developed models with a new set of experimental data and the average percentage deviation of 1.13 % and 3.86 % for attenuation and turbulence strength respectively are obtained. These models are implemented in the MATLAB environment along with the selected models for performing the comparative analysis. The measurement values and prediction results are captured for a diurnal period in different seasons for a one year period. The comparison response of prediction results of all the models and practically measured values are graphically illustrated. The prediction errors are analyzed in terms of RMSE and SAE and graphically illustrated. The greatest prediction accuracy of the developed models in different seasons is highlighted in Chapter 3.

Atmospheric changes and their effects on propagating optical beam wandering are introduced. The results of a literature review are presented. The enhanced optoelectronic experimental setup with suitable sensors to get the beam position information and beam steering optoelectronic control system are explained in Chapter 4. An optical shaker is incorporated at the transmitter assembly to introduce disturbance on both axes as a jitter in addition to the atmospheric disturbances to characterize the frequency of the control that the system could successfully operate. The working principle of OPD and related formulas for estimating the beam centroid information, radial displacement and angular deflection with respect to the direction of beam propagation are detailed. The steady state and calibration response of OPD, piezoelectric actuators and piezo driving amplifier are discussed. The nonlinear behavior of the main devices is highlighted. The formula designed to normalize the experimental data and development of mathematical models to achieve the correction control action using the response surface obtained from the real-time beam stabilization experiment conducted in different seasons are detailed. The control signal's prediction accuracy of the developed models (RSM controllers) is analyzed using the ANOVA tools and finally a full model is finalized. Development of a neural-controller model using the input-output training data sets and investigation of its performance are performed using the testing data sets and the results are presented in Chapter 4. The residual plots and open-loop control responses of both the controllers are illustrated and from them, the better performance of the neural-controller is revealed. The percentage of error of controller signal predicted by the full model controller and neural-controller are analyzed in different trials and the min/max values of 0.2 % and 0.13 % respectively are obtained from the neural-controller. These two controllers are implemented in the MATLAB environment and deployed at the receiver as an autonomous control system to stabilize the beam always on the center of the detector plane. The beam position information is obtained by processing the OPD outputs using MPAC and applied into the controllers in a closed loop control configuration. The performance improvements of the FSO link using the developed beam steering control system are evaluated in terms of beam alignment (compass plot and beam center radial displacement) and stability of the detector maximum output and discussed in Chapter 4.

The need to control operations with low power, compactness and high speed, parallel and pipelined process are introduced. The results of literature reviews are presented. The enhancement of the experimental setup with the hardware (FPGA) control platform and suitable signal conditioning electronics are explained in Chapter 5. The associated estimation formulas are interoperated. The digital design using a Xilinx system generator to implement a data normalization algorithm and RSM controller are explained. The novel pipelined-parallel digital architecture developed in FPGA to implement the neural-controller with various digital modules: clock manager, signal digitization and data processing unit, weight and bias memory management circuit, neuron unit, data routing ring circuit, multiply accumulation unit, and serial clock manager unit are detailed using timing diagram, architecture, design flow graph and data communication data format in Chapter 5. The real-time performances of the developed controllers are investigated in open loop control configuration to validate the control signal prediction accuracy of the controllers in beam stabilization as well as to understand the behavior of the entire system. The percentage of prediction deviations are computed for various new data sets, and the best control performance, as required to mitigate beam wandering, is attained when the neural-controller is used. Therefore, the neural-controller is finally selected as the best candidate controller for predicting the control signal by processing the beam position information to perform the control action and other associated investigations are carried out. The estimated PSD plot illustrates the unique features of the neural-controller in closed loop control configuration. The statistics of beam wandering (beam centroid motion) on the 2D plane of OPD are analyzed with the control on and off conditions and the greatest improvement attained with the control on is discussed. The Effective Scintillation Index (ESI) is estimated using the ensemble average data in a diurnal period for different seasons and the results demonstrated the significance of the developed control system. The normalized impulse response of the FSO link is investigated and the link reliability improvements are analyzed in Chapter 5.

The advantages and features of FSOC over RF in communication perspective are described in Chapter 6. A constructed simplex data transmission experimental setup with beam steering system is explained. Pseudo Random Binary Sequence (PRBS) data pattern generator (data source) at the Asynchronous Transfer Mode (ATM) rate of 155 Mbps is used at the transmitter. The serial data stream is applied to the OOK-IM optical modulator and the information bearing optical wave is permitted to travel in the atmospheric turbulent channel for the data link range of 0.5 km. The OOK-DD demodulation scheme is used at the receiver. The theoretical estimation and practical measurements of various communication quantitative metrics: received power (P_r), transmittance (T), eye pattern, Q-factor, Signal to Noise Ratio (SNR), Bit Error Rate (BER), link margin, and decision threshold values related to the unipolar NonReturn to Zero–On Off Keying–Intensity Modulation–Direct Detection (NRZ-OOK-IM-DD) signaling scheme are presented. The entire performance of the data link is verified using the theoretical models and experimental results. The quality and reliability of the data link are evaluated in beam wandering uncontrolled and controlled environments. The geometrical interpretation of the received signal optical wavefront on the detector plane is analyzed as a function of beam shape and radial distance obtained from the OPD. The effect of atmospheric turbulence on OOK-NRZ and BER performance improvement due to the beam steering system are also investigated experimentally. FSOC data link operation at the ATM rate is demonstrated under different turbulence conditions with a control system to mitigate beam wandering.

7.2 Future work

This research work has been completed the objectives listed in Chapter 1. However, the amount of time and work required to theorize and cover the comprehensive channel characterization and measurements are beyond the scope of this work. The following topics are suggested to extend further the research work reported in this thesis.

High data rate: Perfect coupling of received optical signal to the detector plane is achieved with the developed control system. Simplex data transmission at the ATM rate of 155 Mbps is demonstrated using OOK-NRZ modulation scheme. The full duplex FSOC data link at a greater data rate approximately of greater than 1 Gbps or more has to be tested for different modulation schemes like OOK-RZ, Binary Phase Shift Keying (BPSK), M-array Pulse Amplitude Modulation (M-PAM) and Quadrature Phase Shift Keying (Q-PSK) etc. with the beam steering system for the increased link range (> 2 km) to strengthen the experimental study to understand the maximum data rate that the system could operate at (Tang et al. 2010; Sharma and Chadha 2006). The feasibility and limitations of real FSOC are to be experimentally verified for multiple sources of information, i.e. data, speech, video and etc.

Wavelength specifications: This work is accomplished only with an 850 nm optical source. Examination of wavelength dependence of the FSOC channel becomes significant to verify the suitability of the existing atmospheric transmission window or formulating the new atmospheric transmission window more suitable at the test field. Therefore, the complete performance of the FSOC has to be tested as a function of multiple wavelengths with the fixed FSO communication distance.

Optimization and extension of the control algorithm: The developed neural-controller handled only three actuators used for the beam steering application in this work. This control system has to be extended for more, approximately 23, actuators to incorporate the wavefront distortion correction. Comprehensive analyses have to be carried out on the performance of an extended as well as an optimized neural-controller in wavefront distortion correction with simulation and experimental results.

Coherence degradation: The wavefront global tilt of the received signal is corrected in this work using FSM. The wavefront spatially distorted especially when the data link range is above 2 km. In this case, the Shack–Hartman Wavefront Sensor (SHWS) or suitable image processing techniques along with a Deformable Mirror (DM) are to be incorporated to effectively mitigate atmospheric disturbances (spatial distortion). This technique increases the accuracy of optical wavefront measurement and correction in a closed loop control configuration.

Enhancement in channel modeling: In modeling atmospheric attenuation and turbulence strength (C_n^2), Taguchi's experimental data ANOVA method is used for the formulation of regression models. However, formulations of any complex and nonlinear regression models are beyond the scope of this work. Conducting the experiment in different geographical locations and acquiring the corresponding data becomes significant. In future work, it is recommended to further investigate this topic, in order to provide a more comprehensive model which takes into accounts weather data from different geographical and environmental locations for the estimation of atmospheric attenuation and turbulence strength (C_n^2). The proposed model will definitely yield better prediction than other models in predicting atmospheric attenuation and turbulence strength (C_n^2).

Strong turbulence: The performance of the FSOC has been theoretically and experimentally investigated for very weak to very strong turbulence conditions with an autonomous beam steering system in this book. As mentioned in Chapter 6, the influence of the FSOC channel depends on the strength of the turbulence and the length of the

FSOC link. Turbulence influence increases with the increasing link length. In future work, the length of the FSOC link should be increased in a real-world open atmosphere in order to increase the strength of the turbulence and perform a more comprehensive experimental study. Note that the strength of turbulence also depends on the various environmental factors along the FSOC link (Huichun et al. 2006). This will help to verify log-normal, Gamma-Gamma and negative exponential models and to investigate the BER performance of FSOC under different turbulence conditions. An adaptive threshold adjustment demodulation technique could be used for perfect decision-making in reconstructing the received data.

Spatial diversity: Due to the very strong turbulence, the FSOC signal suffers fading due to multipath propagation, which can be mitigated by spatial diversity techniques such as Multiple Inputs and Multiple Outputs (MIMO) optical antennas that can be equipped at the transmitter and/or receiver sites in order to mitigate the effect of turbulence (Cvijetic et al. 2008; Garcia-Zambrana et al. 2009; Letzepis and Fabregas 2008; Fath and Haas 2013).

Route diversity: FSOC communications link performance is highly affected when propagating through the time-spatially variable atmospheric environment. In Zvanovec et al. (2013) and Libich et al. (2012), route diversity has been adopted in order to improve signal reception under weak turbulence condition. Therefore, route diversity has to be investigated for very weak to very strong atmospheric conditions as a mitigation technique (Garcia-Zambrana et al. 2009).

Hybrid FSOC/RF communication using channel coding: The reliability of an FSO communication system mainly depends on the atmospheric weather conditions. One of the biggest challenges is the attainment of 99.999 % link availability during all weather conditions (Wu & Kavehrad 2007). The hybrid FSOC/RF link combined with the Forward Error Control (FEC) channel coding is one possible option as the RF system could be utilized as the back-up link but at a reduced data rate when turbulence strength is very strong. The channel coding can also improve the overall system reliability (Djordjevic 2010; Sandalidis 2011). Therefore, the performance of the FSOC has to be studied for different modulation techniques with different error control algorithms (Lee et al. 2011).

Radio over FSOC: Transmission of a modulated RF signal using an FSO communications links has been receiving a lot of research interest recently (Dat et al. 2007). The Radio over Free Space Optics Communication (RoFSOC) system has potential to be a cost effective and reliable technology for bridging innovative wireless technologies networks facilities (Kazaura et al. 2008). RoFSOC is a next generation access technology suitable for transmission of heterogeneous wireless service signals especially in areas which lack broadband connectivity. There is a need to conduct initial investigations focusing on studying the deployment environment characteristics, which influence the performance of RoFSOC systems. Therefore, it is recommended to investigate the effects of atmosphere on RoFSOC (Yan et al. 2005).

References

Afonso, A; De Bougrenet de la Tocnaye, JL & Barge, M 2006, 'Dynamic gain and channel equalizers or wavelength blocker using free space: common elements and differences', Journal of Lightwave Technology, vol. 24, no. 3, pp. 1534–1542.

Ahilan, C; Kumanan, S; Sivakumaran, N & Dhas Edwin Raja, J 2013, 'Modeling and prediction of machining quality in CNC turning process using intelligent hybrid decision making tools', Applied Soft Computing, vol. 13, no. 3, pp. 1543–1551.

Ahmed, Yaseen Adnan & Hasegawa, Kazuhika 2013, 'Automatic ship berthing using artificial neural network trained by consistent teaching data using nonlinear programming method', Engineering applications of artificial intelligence, vol. 26, no. 10, pp. 2287–2304.

Ali, Mazin Ali A 2013a, 'Analysis study of rain attenuation on optical communications link', International Journal of Engineering, Business and Enterprise Applications, vol. 13, no. 313, pp. 18–24.

Ali, Mazin Ali A 2013b, 'Atmospheric Turbulence Effect on Free Space Optical Communications', International Journal of Emerging Technologies in Computational and Applied Sciences, vol. 13-364, pp. 345–351.

Ali, Mazin Ali A 2014, 'Free Space Optical wireless communications under turbulence channel effect', IOSR Journal of Electronics and Communication Engineering, vol. 9, no. 3, pp. 1–8.

Anbuselvi, T & Bazil Raj, A 2014, 'Design and Verification of Pipelined Parallel Architecture Implementation in FPGA for Bit Error Rate Tester', International Journal of Research in Engineering and Technology, vol. 3, spl. iss. 2, pp. 28–34.

Andrews, Larry C; Phillips, Ronald L & Hopen, Cynthia Y 2001, 'Laser Beam Scintillation with Applications', SPIE, USA.

Andria, G; Cavone, G; Di Lecce, V & Lanzolla, AML 2005, 'Model Characterization in Measurements of Environmental Pollutants via Data Correlation of Sensor Outputs', IEEE Transactions on Instrumentation and Measurement, vol. 54, no. 3, pp. 1061–1066.

Arnon, S; Rotman, SR & Kopeika, NS 1999, 'Bandwidth maximization for satellite laser communications', IEEE transactions on aerospace and electronic systems, vol. 35, no. 2, pp. 675–681.

Arnon, S & Kopeika, NS 1997, 'Adaptive optical transmitter and receiver for space communication through thin clouds', Applied Optics, vol. 36, no. 9, pp. 1987–1993.

Arnon, Shlomi 2003, 'Optimization of Urban Optical Wireless Communication Systems', IEEE Transactions on Wireless Communications, vol. 2, no. 4, pp. 626–629.

Asokan, P; Ravikumar, R; Jeyapaul, R & Santhi, M 2008, 'Development of multi-objective optimization models for electrochemical maching process', International Journal on Advanced Manufacturing Technology, vol. 39, no. 1–2, pp. 55–63.

Awan, MS; Brandl, P; Leitgeb, E; Nadeem, F; Csugai-Horvath, L & Nebuloni, R 2009a, 'Transmission of high data rate optical signals in fog and Snow conditions', Proceedings of the 1st International Conference on Wireless Communication, Vehicular Technology, Information Theory and Aerospace & Electronic Systems Technology, Aalborg, pp. 702–706.

Awan, MS; Marzuki; Leitgeb, E; Hillibrand, B; Nadeem, F & Khan, MS 2009b, 'Cloud attenuations for free-space optical links', Proceedings of international workshop on satellite and space communications, Tuscany, pp. 274–278.

Awan, MS; Horwath, LC; Muhammad, SS; Leitgeb, E; Nadeem, F & Khan, MS 2009c, 'Characterization of Fog and Snow Attenuations for Free Space Optical Propagation', Journal of Communications, vol. 4, no. 8, pp. 533–545.

Bade, Stephen L & Hutchings, Brad L 1994, 'FPGA based stochastic neural networks – implementation', Proceedings of the IEEE workshop on FPGAs for Custom Computing, Napa, CA, pp. 189–198.

Banuelos-Saucedo, MA; Castillo-Hernández, J; Quintana-Thierry, S; Damián-Zamacona, R; Valeriano-Assem, J; Cervantes, RE; Fuentes-González, R; Calva-Olmos, G & Pérez-Silva, JL 2003, 'Implementation of a neuron model using FPGAS', Journal of Applied Research and Technology, vol. 1, no. 03, pp. 248–255.

Bao, Lianwei; Zhao, Hengkai; Zheng, Guoxing & Ren, Xinping 2012, 'Scintillation of THz transmission by atmospheric turbulence near the ground', Proceedings of 5th International conference on Advanced computational intelligence, china, pp. 932–936.

Barrios, R & Dios, F 2012, 'Wireless optical communications through the turbulent atmosphere: a review', Das, N (Ed.), Optical communications systems, InTech, pp. 3–40.

Barry, J Dale 1984, 'Design and system requirements imposed by the selection of GaAs/GaAlAs single mode laser diodes for free space optical communications', IEEE journal of quantum electronics, vol. 20, no. 5, pp. 478–491.

Barsimantov, Ohad & Nikulin, Vladimir V 2011, 'Adaptive Optimization of a Free Space Laser Communication System under Dynamic Link Attenuation', Journal of Optical Communication and Networking, vol. 3, no. 3, pp. 215–222.

Bataille, P 1992, 'Analysis of the behavior of a telecommunications system operated optical 0.83 microns in the lower atmosphere', Ph.D Thesis, University of Rennes, France.

Bazil Raj, A et al. 2010, 'Terrestrial Free Space Line of Sight Optical Communication (TFSLSOC) using Adaptive Control Steering System with Laser Beam Aligning, Tracking and Positioning (ATP)', IEEE Int., Conf. Wireless Communication & Sensor Computing, Chennai, pp. 1–5.

Bazil Raj, A & Vijaya Selvi, A 2011, 'Lower-Order Adaptive Beam Steering System in Terrestrial Free Space Point-to-Point Laser Communication using Fine Tracking Sensor', IEEEInt. Conf., Signal Processing, Communication, Computing and Networking Technologies, Thuckalay, pp. 699–704.

Bazil Raj, A et al. 2012, 'Low Cost Beam Steering System for FSOC to SMF Coupling', IEEE Int. Conf., Advances in Engineering, Science and Management, Nagapattinam, pp. 49–54.

Bazil Raj, A et al. 2014a, 'Real-time Measurement of Meteorological Parameters for Estimating Low Altitude Atmospheric Turbulence Strength (Cn2)', IET- Science, Measurement & Technology, vol. 8, iss. 6, pp. 459–469.

Bazil Raj, A et al. 2014b, 'Mitigation of Beam Fluctuation due to Atmospheric Turbulence and Prediction of Control Quality using Intelligent Decision Making Tools', Applied optics, vol. 53, no. 17, pp. 37962–3806.

Bazil Raj, A et al. 2014c, 'A Direct and Neural Controller Performance Study with Beam Wandering Mitigation Control in Free Space Optical Link', Optical Memory and Neural Networks (Information Optics), vol. 23, iss. 3, pp. 111–129.

Bazil Raj, A et al. 2014d, 'Intensity Feedback based Beam Wandering Mitigation in Free Space Optical Communication using Neural Control Technique', EURASIP Journal on Wireless Communications and Networking, DOI: 10.1186/1687-1499-2014-160, pp. 1–18.

Bazil Raj, A & Vijaya Selvi, A 2015a, 'Comparison of Different Models for Ground-Level Atmospheric Attenuation Prediction with New Model According to Local Weather Data for FSO Applications', Journal of Optical Communication, DOI: 10.1515/joc-2014-0054, pp. 1–16.

Bazil Raj, A et al. 2015b, 'Comparison of Different Models for Ground-Level Turbulence Strength (C_n^2) Prediction with New Model According to Local Weather Data for FSO Applications', Applied Optics, vol. 54, iss. 4, pp. 802–815.

Bazil Raj, A et al. 2015c, 'Design of Cognitive Decision Making Controller for Autonomous Online Adaptive Beam Steering in Free Space Optical Communication System', Wireless Personal Communications, vol. 84, pp. 765–799.

Bell, AG 1880, 'On the production and reproduction of sound by light', American Journal of Sciences, vol. 3, pp. 305–324.

Belmonte, Aniceto & Kahn, Joseph M 2009, 'Capacity of coherent free-space optical links using atmospheric compensation techniques', Optics Express, vol. 17, no. 4, pp. 2763–2773.

Bendersky, Sergey; Kopeika, Norman S & Blaunstein, Natan 2004, 'Atmospheric optical turbulence over land in Middle East coastal environments: prediction modeling and measurements', Applied Optics, vol. 43, no. 20, pp. 4070–4079.

Benrekia, F & Attari, M 2010, 'A Floating Point Multiplier based FPGA Synthesis for Neural Networks Enhancement', International Journal of Engineering Science and Technology, vol. 2, no. 5, pp. 1433–1440.

Biswas, A; Hemmati, H & Lesh JR 1999, 'High Data-rate Laser Transmitters for Free-Space Laser Communications', Proceedings of SPIE 3615, Free-Space Laser Communication Technologies XI, San Jose, CA , pp. 269–277.

Bloom, Scott; Korevaar, Eric; Schuster, John & Willebrand, Heinz 2003, 'Understanding the Performance of Free-Space Optics [Invited]', Journal of Optical Networking, vol. 2, no. 6, pp. 178–199.

Boffi, Pierpaolo; Piccinin, David; Mottarella, Diego & Martinelli, Mario 2000, 'All-optical free-space processing for optical communication signals', Optics Communications, vol. 181, iss. 1-3, pp. 79–88.

Bouchet, Oliver; Sizun, Herve; Boilsrobert, Christian & de Fornel, Frédérique 2006, 'Free Space Optics: Propagation and Communication', ISTE, UK.

Brandenburg, Jacob C & Liu, John Q 2009, 'Signal detection for optical communications through the turbulent atmosphere', IEEE Transactions on Communication, vol. 57, no. 11, pp. 3425–3432.

Carruthers, Jeffrey B & Kahn, Joseph M 2000, 'Angle diversity for nondirected wireless infrared communication', IEEE Transactions on Communications, vol. 48, no. 6, pp. 960–969.

Chadha, Devi 2005, 'Feasibility study of free space optical communication link through atmospheric turbulent channel', Proceedings of conference on Optoelectronic and Microelectronic Materials and Devices, Brisbane, Qld, pp. 311–314.

Chand, Naresh 2000, 'Current progress and challenges in bringing fiber to home', Proceedings of International conference on fiber optics and photonics, Kharagpur, pp. 167.

Chang, Mark PJL; Roura, Erick A; Font, Carlos O; Gilbreath, Charmaine & Oh, Eun 2006, 'Applying the Hilbert-Huang decomposition to horizontal light propagation Cn2 data', Proceedings of SPIE 6268, Advances in Stellar Interferometry, Orlando, Florida, pp. 62683E.

Cheever, Erik 2010, 'Serial Communications tutorial' – Available from: http://www.swarthmore.edu/NatSci/echeeve1/Class/e91/Lectures/E91%2810%29Serial.pdf

Chen, Yen-Hsiang; Hu, Kai-Ti & Ruan, Shanq-Jang 2012, 'Statistical skin color detection method without color transformation for real-time surveillance systems', Engineering Applications of Artificial Intelligence, vol. 25, iss. 7, 1331–1337.

Chow, CW; Yeh, CH; Liu, YF & Huang, PY 2013, 'Mitigation of Optical Background Noise in Light Emitting Diode (LED) Optical Wireless Communication Systems', IEEE Photonics Journal, vol. 5, no. 1, pp. 7900307-1–7.

Ciaramella, E; Arimoto, Y; Contestabile, G; Presi, M; D'Errico, A; Guarino, V & Matsumoto, M 2009, '1.28 Terabit/s (32 × 40 Gbit/s) WDM Transmission System for Free Space Optical Communications', IEEE Journal on Selected Areas in Communications, vol. 27, no. 9, pp. 1639–1645.

Colvero, CP; Cordeiro, MCR & von der Weid, JP 2007, 'FSO systems: rain, drizzle, fog and haze attenuation at different optical windows propagation', Proceedings of SBMO/IEEE MTT-S International Microwave Optoelectronics Conference, pp. 563–568.

Coquilla, Rachael V 2010, Anemometer Calibration for Wind Energy Applications. Available from: http://otechwind.com/wp-content/uploads/MSC-2010-Paper.pdf

Corrigan, Paul; Martini, Rainer; Whittaker, Edward A & Bethea, Clyde 2009, 'Quantum cascade lasers and the kruse model in free space optical communication', Optics Express, vol. 17, no. 6, pp. 4355–4359.

Cowley, William G; Lechner, Gottfried; Biggs, Colin & Grant, Kenneth J 2011, 'Transceiver design for reliable free space optical communications', Proceedings of Australian Communications Theory Workshop (AusCTW), Melbourne, VIC, pp. 94–99.

Cox, MG & Harris, PM 2006, 'Software Support for Metrology Best Practice Guide No. 6 Uncertainty Evaluation', NPL Report, DEM-ES-011, ISSN 1744-0475, pp. 1–175.

Cvijetic, N; Dayou, Q; Jianjun, Y; Yue-Kai, H & Ting, W 2009, '100 Gb/s perchannel free-space optical transmission with coherent detection and MIMO processing,' Proceedings of 35th European Conference on Optical Communication, pp. 1–2.

Cvijetic, N; Wilson, SG & Brandt-Pearce, M 2008, 'Performance bounds for free-space optical MIMO systems with APD receivers in atmospheric turbulence', IEEE Journal on Selected Areas in Communications, vol. 26, no. 3, pp. 3–12.

Daalen, Max van, Jeavons, Peter & Shawe-Taylor, John 1997, 'A stochastic neural architecture that exploits dynamically reconfigurable FPGAs', Proceedings of the IEEE workshop on FPGAs for custom computing machines, Napa, CA, pp. 202–211.

Das, Narottam 2012, Optical Communications Systems, INTECH, Croatia. Available from INTECH Books [3rd October 2012].

Dat, Pham Tien; Naila, Chedlia Ben; Liu, Peng; Wakamori, Kazuhiko; Matsumoto, Mitsuji & Tsukamoto, Katsutoshi 2011, 'Next Generation Free Space Optics Systems for Ubiquitous Communications', PIERS Online, vol. 7, no. 1, pp. 75–80.

Dat, Pham Tien; Shah, AM; Kazaura, K; Wakamori, K; Suzuki, T; Omae, K; Matsumoto, M; Aburakawa, Y; Takahashi, K; Nakamura, T; Higashino, T; Tsukamoto, K & Komaki, S 2007, 'Investigation of suitability of RF signal transmission over FSO links', Proceedings of international symposium on high capacity optical networks and enabling technologies, Dubai, pp. 1–6.

Tianping Deng; Yimin Lu & Xiaoxiao 2005, 'Performance evaluation of free space optical communication system', Proceedings of International Conference on Wireless Communications, Networking and Mobile Computing, pp. 587–590. DOI: 10.1109/WCNM.2005.1544112

Dettmer, R 2001, 'A ray of light, free space optical transmission', IEE Review, vol. 47, pp. 32–33.

Dias, Fernando Morgado; Antunes, Ana & Mota, Alexandre Manuel 2004, 'Artificial neural networks: a review of commercial hardware', Engineering Applications of Artificial Intelligence, vol. 17, no. 8, pp. 945–952.

Djordjevic, Ivan B 2007, 'LDPC-coded MIMO optical communication over the atmospheric turbulence channel using Q-ary pulse-position modulation', Optics Express, vol. 15, no. 16, pp. 10026–10032.

Djordjevic, Ivan B 2010, 'Adaptive modulation and coding for Free-space optical channels', Journal of Optical Communication Network, vol. 2, no. 5, pp. 221–229.

Djordjevic, Ivan B 2011, 'Heterogeneous Transparent Optical Networking based on Coded OAM Modulation', IEEE Photonics Journal, vol. 3, no. 3, pp. 531–537.

Doss-Hammel, Steve; Oh, Eun; Ricklin, Jennifer; Eaton, Frank; Gilbreath, Charmaine & Tsintikidis, Dimitri 2004, 'A Comparison of Optical Turbulence Model', Proceedings of SPIE, Free-Space Laser Communications IV, vol. 5550, pp. 236–245.

Dror, Itai; Atar, S; Grossman, Shlomit & Kopeika, Norman S 1996, 'Accurate method for prediction of atmospheric transmission according to weather', Optical Engineering, vol. 35, no. 9, pp. 2548–2555.

Efe, Murat; Ali, Fiket; Ozek, Faruk; Atherton, Derek P & Ali, Falah 2003, 'Double ended optical beam tracking', Proceedings of 2003 IEEE Conference on Control Applications, pp. 1221–1226.

Elkamchouchi, Hassan M & El-Shimy, Mohamed A 2006, 'Monte carlo simulation of light propagation through the troposphere for free space optical communications', Proceedings of 23rd National radio science conference, Egypt, pp. 1–8.

Esposito, Joel M; Barton, Owen & Kohler, Joshua 2011, Tutorial: serial communications in Matlab. Available from: http://www.usna.edu/Users/weapsys/esposito/roomba.matlab/SerialCommunicationinMatlabV2.doc

Fares, Salma Ait & Adachi, Fumiyuki 2010, Mobile and Wireless Communications Network Layer and Circuit Level Design, INTECH, Croatia.

Farmahini-Farahani, Amin; Vakili, Shervin; Fakhraie, Sied Mehdi; Safari, Saeed & Luca, Caro 2010, 'Parallel scalable hardware implementation of asynchronous discrete particle swarm optimization', Engineering Applications of Artificial Intelligence, vol. 23, no. 2, pp. 177–187.

Fasinmirin, JT; Oguntunde, PG; Ladipo, KO & Dalbianco, L 2011, 'Development and calibration of a self-recording cup anemometer for wind speed measurement', African Journal of Environmental Science and Technology, vol. 5, no. 3, pp. 212–217.

Fath, T & Haas, H 2013, 'Performance Comparison of MIMO Techniques for Optical Wireless Communications in Indoor Environments'.

Ferdinandov, E; Pachedjiveva, B; Bonev, B & Saparev, SL 2007, 'Joint influence of heterogenous stochastic factors on bit-error rate of ground-to-ground free-space laser communication system', Optics communications, vol. 270, iss. 2, pp. 121–127.

Ficocelli, Maurizio & Amara, Foued Ben 2012, 'Online tuning of retinal imaging adaptive optics', IEEE transactions on Control Systems Technology, vol. 30, no. 3, pp. 747–754.

Fidler, Franz; Knapek, Markus; Horwath, Joachim & Leep, Walter R 2010, 'Optical Communications for High-Altitude Platforms', IEEE Journal of Selected Topics in Quantum Electronics, vol. 16, no. 5, pp. 1058–1070.

Fielhauer, KB; Boone, BG; Bruzzi, JR; Kluga, BE; Connelly, JR; Bierbaum, MM; Gorman, JJ & Dagalakis, N 2003, 'Comparison of Macro – Tip/Tilt and Meso – Scale Position Beam Steering Transducers for a Free Space Optical Communications using a Quadrant Photodiode Sensor', Proceedings of SPIE 5160, Free – Space Laser Communication and Active Laser Illumination III, San Diego, California, USA, pp. 192–203.

Fischer, KW; Witiw, MR & Baars, JA 2004, 'Atmospheric Laser Communication', Bulletin of the American Meteorological Society, vol. 85, pp. 725–732.

Fletcher, G; Hicks, T & Laurent, B 1991, 'The SILEX optical interorbit link experiment', Electronics & Communication Engineering Journal, vol. 3, pp. 273–279.

Font, Carlos O; Chang, Mark PJL; Oh, Eun & Gilbreath, Charmaine 2006, 'Humidity Contribution to the Refractive Index Structure Function C_n^2', Proceedings of SPIE 6215, Atmospheric Propagation III, Orlando (Kissimmee), FL, pp. 1–9.

Forin, DM; Incerti, G; Beleffi, GT; Teixeira, A; Costa, L; Andrè, PDB; Geiger, B; Leitgeb, E & Nadeem, F 2010, Free Space Optical Technologies, IN-TECH,China.

Frederickson, Paul; Davidson, Kenneth: Zeisse, Carl & Bendall, Ike 1998, 'Near-Surface Scintillation (Cn2) Estimates from a Buoy using Bulk Methods during EOPACE', Proceedings of RTO SET symposium on 'E-O Propagation, Signature and System Performance Under Adverse Meteorological Conditions Considering Out-of-Area Operations, Italy, pp. 16-1–16-9.

Freude, Wolfgang; Schmogrow, Rene; Nebendah, Bernd; Winter, Marcus; Josten, Arne; Hillerkuss, David; Koenig, Swen; Meyer, Joachim; Dreschmann, Michael; Huebner, Micheal; Koss, Christina; Becker, Juergen & Leuthold, Juerg 2012, 'Quality metrics for optical signals. Eye Diagram, Q-factor, OSNR, EVM and BER', Proceedings of 14th International Conference on Transparent Optical Networks, Coventry, pp. 1–4.

Gago, F; Rodrfguez-Ramos, L; Herrera, G; Gigante, J; Alonso, A; Viera, T; Piqueras, J & Diaz, JJ 2007, 'Tip-Tilt Mirror Control based on FPGA for an Adaptive Optics System', Proceedings of 3rd Southern Conference on Programmable Logic, Mar del Plata, pp. 19–26.

Gappmair, W 2011, 'Further results on the capacity of free-space optical channels in turbulent atmosphere', IET Communication, vol. 5, no. 9, pp. 1262–1267.

Gappmair, W & Flohberger, M 2009, 'Error performance of coded FSO links in turbulent atmosphere modeled by gamma-gamma distributions', IEEE Transactions on Wireless Communications, vol. 8, pp. 2209–2213.

Garcia-Zambrana, A; Castillo-Vazquez, C; Castillo-Vazquez, B & Hiniesta-Gomez, A 2009, 'Selection transmit diversity for FSO links over strong atmospheric turbulence channels', IEEE Photonics Technology Letters, vol. 21, pp. 1017–1019.

Garlington, Tom & Babbitt, Maj Joel 2005, 'Analysis of free space optics as transmission technology', U.S. Army Information Systems Engineering Command. no. 3. pp. 1–11. Available from: http://web.archive.org/web/20070613000248/http://www.hqisec.army.mil/isec/publications/Analysis_of_Free_Space_Optics_as_a_Transmission_Technology_Mar05.pdf

Gebhart, Michael; Leitgeb, Erich; Naboulsi, Maher Al; Sizun, Herve & de Fornel, Frederique 2004, 'Measurements of light attenuation at different wavelengths in dense fog conditions for FSO applications, COST 270 Short term scientific mission7 report', University of Bourgogne Dijon, France, pp. 1–35.

Ghaseemlooy, Z; Poopola, W & Rajbhandri, S 2012, 'Optical wireless communications, system and channel modelling with matlab', CRC Press, London, UK.

Ghassemlooy, Z & Popoola, WO 2010, 'Mobile and Wireless Communications Network Layer and Circuit Level Design', INTECH, Croatia.

Ghassemlooy, Zabih; Minh, Hoale; Rajbhandari, Sujan; Perez, Joaguin & Ijaz, Muhammad 2012, 'Performance Analysis of Ethernet/ Fast-Ethernet Free Space Optical Communications in a Controlled Weak Turbulence Condition', Journal of Lightwave Technology, vol. 30, no. 13, pp. 2188–2194.

Ghosh, S; Basu, K & Das, SK 2005, 'An architecture for next-generation radio access networks', IEEE Network, vol. 19, pp. 35–42.

Giggenbach, Dirk & Horwath, Joachim 2005, 'Optical free space communications downlinks from stratospheric platforms – overview on STROPEX, the optical communications experiment CAPANINA', Proceedings of the IST Mobile and Wireless Communications summit, Dresden, pp. 1–4. Available from: http://www.eurasip.org/Proceedings/Ext/IST05/papers/483.pdf

Gomperts, Alexander; Ukil, Abisek & Zurfluh, Franz 2011, 'Development and Implementation of Parameterized FPGA-Based General Purpose Neural Networks for Online Applications', IEEE Transactions on Industrial Informatics, vol. 7, no. 1, pp. 78–89.

Goodwin, FE 1970, 'A review of operational laser communication systems', Proceedings of the IEEE, vol. 58, pp. 1746–1752.

Grabner, Martin & Kvicera, Vaclav 2007, 'On the relation between atmospheric visibility and optical wave attenuation', Proceedings of 16th IST Mobile and Wireless Communications Summit, Budapest, pp. 1–5.

Grabner, Martin & Kvicera, Vaclav 2008, 'Experimental study of atmospheric visibility and optical wave attenuation for free-space optics communications', Czech Science Foundation project no. 102/08/0851. Available from: http://ursifrance.institut-telecom.fr/pages/pages_ursi/URSIGA08/papers/F06p5.pdf

Grabner, Martin & Kvicera, Vaclav 2011, 'The wavelength dependent model of extinction in fog and haze for free space optical communication', Optics Express, vol. 19, no. 4, pp. 3379–3386.

Grant, Kenneth J; Moore, Christopher I; Burris, Harris R; Mahon, Rita; Wasiezko, Linda; Rabinovich, William; Swingen, Lee; Ferraro, Mike; Suite, Michelle; Stell, Mena: Gilbreath, Charmaine; Clare, Bradley A & Corbett, Kery A 2006a, 'Mitigation of scintillation noise in a 32-km maritime path', Proceedings of Conference on Lasers and Electro-Optics, and 2006 Quantum Electronics and Laser Science, Long Beach, CA, pp. 1–2.

Grant, Kenneth J; Murphy, James; Mahon, Rita; Burris, Harris R; Rabinovich, William S; Moore, Christopher I; Wasiczko, Linda M; Goetz, Peter G; Suite, Michele; Ferraro, Mike S; Gilbreath, G

Charmaine; Clare, Bradley A; Mudge, Kerry A & Chaffey, Jason 2006b, 'Free space optical transmission of AM composite video signals using InGaAS modulating retro-reflectors' Proceedings of Conference on Optoelectronic and Microelectronic Materials and Devices, Perth, WA, pp. 200–203.

Gu, Guo Ying & Zhu, LM 2010, 'High-speed tracking control of piezoelectric actuators using an ellipse-based hysteresis model', Review of Scientific Instruments, vol. 81, no. 8, pp. 1–9.

Gupta, Sudhir Kumar; Mathew, Shibu K & Venkatakrishnan, P 2006, 'Development of Solar Scintillometer', Journal of Astrophysics, vol. 27, no. 2-3, pp. 315–320.

Gupta, Vikas; Khare, K & Singh, RP 2009, 'FPGA design and implementation issues of artificial neural network based PID controllers', Proceedings of International conference on Advances in Recent technologies in communication and computing, Kottayam, Kerala, pp. 860–862.

Hagan, Martin T & Demuth, Howard B 1999, 'Neural Networks for Control, Invited Tutorial', Proceedings of American Control Conference, San Diego, CA, pp. 1642–1656.

Hashmi, Ali Javed; Eftekhar, Ali Asghar; Adibi, Ali & Amoozegar, Farid 2010, 'Analysis of telescope array receivers for deep-space inter-planetary optical communication link between Earth and Mars', Optics Communications, vol. 283, no. 10, pp. 2032–2042.

Hemmati, H 2006, 'Deep space optical communications', Wiley-Interscience, California.

Hemmati, Hamid 2009, 'Near-Earth Laser Communication', CRC Press, Newyork, pp. 59–96.

Hsu, Chun-Fei 2008, 'Adaptive growing-and-pruning neural network controller for a linear piezoelectric ceramic motor', Engineering Applications of artificial intelligence, vol. 21, no. 8, pp. 1153–1163.

Hu, Guo-Yong; Chen, Chang-Ying & Chen, Zhen-Qiang 2007, 'Free space optical communication using visible light', Journal of Zhejiang University Science A, vol. 8, no. 2, pp. 186–191.

Huichun, Yang; Wentao, Zhang & Xing, Jiang 2006, 'The analysis and research for the laser beam through random atmosphere channel', Proceedings of International Conference on Communications, Circuits and Systems, Guilin, pp. 1892–1896.

Hunt, JCR 1999, 'Environmental Forecasting and Turbulence Modeling', Physica D, vol. 133, no. 1-4, pp. 270–295.

Hurh, Yoon-Suk; Shin, Kyun-Woon; Lee, Sang-Hoon & Lee, Jae-Seung 2005, 'Weather-Insensitive Optical free-space communication using gain-saturated optical fiber amplifiers', Journal of Lightwave technology, vol. 23, no. 12, pp. 4022–4023.

Hutt, Daniel L & Tofsted, David H 2000, 'Effect of atmospheric turbulence on propagation of ultraviolet radiation', Optics and Laser technology, vol. 32, pp. 39–48.

Ijaz, M; Ghassemlooy, Z; Le Minh, H; Rajbhandari, S; Perez, J & Gholami, A 2011, 'Bit error rate measurement of free space optical communication links under laboratory-controlled fog conditions', Proceedings of 16th European Conference on Networks and Optical Communications (NOC), Newcastle-Upon-Tyne, pp. 52–55.

Ijaz, Muhammad; Ghassemlooy, Zabih; Perez, Joaquin & Brazda, Vladimir 2013b, 'Enhancing The Atmospheric Visibility and Fog Attenuation using a Controlled FSO Channel', IEEE Photonics Technology Letters, vol. 25, no. 13, pp. 1262–1265.

Ijaz, Muhammad; Ghassemlooy, Zabih; Pesek, Jiri; Fiser, Ondrej; Minh, Hoa Le & Bentley, Edward 2013a, 'Modeling of Fog and Smoke Attenuation in Free Space Optical Communications Link Under Controlled Laboratory Conditions', Journal of Lightwave Technology, vol. 31, no. 11, pp. 1720–1726.

Jiang, Yijun; Ma, Jing; Tan, Liyung; Yu, Siyuan & Du, Wenhe 2008, 'Measurement of Optical Intensity Fluctuation over an 11.8 Km Turbulent Path', Optics Express, vol. 16, no. 10, pp. 6963–6973.

Jooshesh, A; Gulliver, TA & Uysal, S 2012, 'Space to ground optical power Transmission', 7th International Conference on Broadband, Wireless Computing, Communication and Applications, Victoria, BC, pp. 302–307.

Juarez, JC; Dwivedi, A; Mammons, A; Jones, SD; Weerackody, V & Nichols, RA 2006, 'Free-space optical communications for next-generation military networks', IEEE Communications Magazine, vol. 44, no. 11, pp. 46–51.
Jurado Navas, A; Gracia Zambrana, A & Puerta Notario, A 2006a, 'Efficient channel model for free space optical communications', Proceedings of IEEE Mediterranean Electrotechnical Conference, Malaga, pp. 631–634.
Jurado Navas, A; Gracia Zambrana, A & Puerta Notario, A 2006b, 'Wind influence on rate-adaptive transmission techniques in free space optical communication', Proceedings of the IEEE International Conference on Communications, Istanbul, pp. 373–378.
Kashani, FD; Reza Hedayati Rad, M; Kazemian, E & Golomohammady, Sh 2013, 'Reliability analysis of an auto-tracked FSO link under adverse weather condition', Optik - International Journal for Light and Electron Optics, vol. 124, iss. 22, pp. 5462–5467.
Katsis, A; Nistazakis, HE & Tombras, GS 2009, 'Bayesian and frequentist estimation of the performance of free space optical channels under weak turbulence conditions', Journal of the Franklin Institute, vol. 346, no. 4, pp. 315–327.
Kaushik, M; Kaushik, AK & Kaushik, S 2012, 'An overview of Technical aspect for Mobile Network Technology', International Journal of Advances in Engineering Science and Technology, vol. 1, no. 2, pp. 97–103.
Kazaura, K; Dat, P; Shah, A; Suzuki, T; Wakamori, K; Matsumoto, M; Higashino, T; Tsukamoto, K & Komaki, S 2008, 'Studies on a next generation access technology using radio over free-space optic links', Proceedings of second international conference on Next Generation Mobile Applications, Services and Technologies, Cardiff, pp. 317–324.
Kazem, Gailan Asad 2013, 'Study the effects of Atmospheric Turbulence on Propagation of Laser Beam', Diyala Journal for Pure Sciences, vol. 9, no. 3, pp. 34–46.
Kedar, Debbie & Arnon, Shlomi 2005, 'Backscattering induced cross talk in WDM optical wireless communication', Journal of Lightwave Technology, vol. 23, no. 6, pp. 2023–2030.
Kepa, K; Coburn, D; Dainty, JC & Morgan, F 2008, 'High Speed Optical Wavefront Sensing with Low Cost FPGAs', Measurement Science Review, vol. 8, no. 4, pp. 87–93.
Khalil, Rafid Ahmed & Al-Kazzaz, Sa'ad Ahmed 2007, 'Digital hardware implementation of artificial neural neurons models using FPGA', Al-Rafidain Engineering, vol. 17, no. 2, pp. 12–24.
Khan, MS; Grabner, M; Muhammad, SS; Awan, MS; Leitgeb, E; Kvicera, V & Nebuloni, R 2012, 'Empirical relations for optical attenuation prediction from liquid water content of fog', Radio engineering, vol. 21, no3, pp. 911–916.
Killinger, D 2002, 'Free space optics for laser communication through the air', Optics and Photonics News, vol. 13, pp. 36–42.
Kim, Isaac I; McArthur, Bruce & Korevaar, Eric 2001, 'Comparison of laser beam propagation at 785nm and 1550nm in fog and haze for optical wireless communications', Proceedings of SPIE 4214, Optical Wireless Communications III, Boston, MA, pp. 26–37.
Konesky, Gregory 2006, 'Application of Adaptive Optics to a Moon-to-Earth Optical Data Link', Proceedings of IEEE Aerospace Conference, Big Sky, MT, pp. 1–7.
Kumar, Naresh & Rana, Ashwani Kumar 2013, 'Impact of various parameters on the performance of free space optics communication system', Optik - International Journal for Light and Electron Optics, vol. 124, no. 22, pp. 5774–5776.
Kumar, Navin; Lourenco, Nuno; Spiez, Michal & Aguiar, Rui L 2008, 'Visible light communication systems conception and VIDAS', IETE technical review, vol. 25, no. 6, pp. 359–367.
Kusnerova, M; Valicekl, J; Harnicarova, M; Hryniewicz, T; Rokosz, K; Palkova, Z; Vaclavik, V; Repka, M & Bendova, M 2013, 'A Proposal for Simplifying the Method of Evaluation of Uncertainities in Measurement Results', Measurement Science Review, vol. 13, no. 1, pp. 1–6.

Le, Quy Ngoc & Jae-Wook, Jeon 2010, 'Neural-network-based low-speed-damping controller for stepper motor with an FPGA', IEEE Transactions of Industrial Electronics, vol. 57, no. 9, pp. 3167–3180.

Leclerc, Troy T; Philips, Ronald L; Andrews, Larry C; Wayne, David T; Saucer, Paul & Crabbs, Robert 2010, 'Prediction of the ground-level refractive index structure parameter from the measurement of atmospheric conditions', Proceedings of SPIE 7685, Atmospheric Propagation VII, Orlando, Florida, USA, pp. 76850A-76858.

Lee, IE: Ghassemlooy, Z; Ng, WP & Rajbhandari, S 2011, 'Fundamental analysis of hybrid free space optical and radio frequency communication systems', Proceedings of Annual Post Graduate Symposium on the Convergence of Telecommunications, Networking and Broadcasting (PGNet2011), Liverpool , pp. 1–5.

Lee, Shinhak; Ortiz Gerry G; Alexander, James W; Portillo, Angel & Jeppesen, Christian 2001, 'Accelerometer-Assisted Tracking and Pointing for Deep Space Optical Communications: Concept, Analysis and Implementations', Proceedings of IEEE Aerospace Conference, vol. 3, Big Sky, MT, pp. 1559–1564.

Lee, Shinhak & Ortiz, Gerry G 2003, 'Inertial sensor assisted acquisition, tracking and pointing for high data rate free space optical communications', Proceedings of 16th Annual meeting of the IEEE Lasers and Electro-Optics Society (LEOS), pp. 87–88.

Leeb, Walter R 2000, 'Space Laser Communications: Systems, Technologies and Applications', Review of Laser Engineering, vol. 28, no. 12, pp. 804–808.

Leibowitz, Brian S; Boser, Bernhard E & Pister, Kristofer SJ 2005, 'A 256-element CMOS imaging receiver for free-space optical communication', IEEE journal of solid-state circuits, vol. 40, no. 9, pp. 1948–1956.

Leitgeb, E; Awan, MS; Brandl, P; Plank, T; Capsoni, C; Nebuloni, R; Javornik, T; Kandus, G; Muhammad, S; Ghassemlooy, F; Loschnigg, M & Nadeem, F 2009, 'Current optical technologies for wireless access', Proceedings of 10th International Conference on Telecommunications, Zagreb, pp. 7–17.

Letzepis, N & Fabregas, A 2008, 'Outage probability of the MIMO Gaussian free-space optical channel with PPM', IEEE International Symposium on Information Theory, pp. 2649–2653.

Li, Jing & Uysal, Murat 2003, 'Achievable information rate for outdoor free space optical communication with intensity modulation and direct detection', Proceedings of Global Telecommunications Conference, pp. 2654–2658.

Libich, J; Zvanovec, S & Mudroch, M 2012, 'Mitgation of time-spatial influence in free-space optical networks utilzing route diversity', Proceedings of SPIE, Free-Space Laser Communication Technologies XXIV, San Francisco, California, USA, pp. 824600.

Liu, P; Kazaura, K; Wakamori, K & Matsumoto, M 2010, 'Studies on Cn2 and its effects on Free Space Optical Communication Systems', 8th Asia-Pacific Symposium on Information and Telecommunication Technologies (APSITT), Kuching, pp. 1–6.

Liu, Wei; Shi, Wenxiao; Cao, Jingtai; Lv, Yaowen; Yao, Kainan; Wang, Shuai; Wang, Jihong & Chi, Xuefen 2014, 'Bit error rate analysis with real-time pointing errors correction in free space optical communication systems', Optik – International Journal for Light and Electron Optics, vol. 125, no. 1, pp. 324–328.

Liu, Xian 2009, 'Free – Space Optics Optimization Models for Building Sway and Atmospheric Interference using Variable Wavelength', IEEE Transactions on Communications, vol. 57, no. 2, pp. 492–498.

Liu Yu Tai; Chen, Neil & Gibson, Steve 2005, 'Adaptive filtering and control for wavefront reconstruction and jitter control in adaptive optics', Proceedings of American Control Conference, Portland, USA, pp. 2608–2612.

Lu, Yafei; Fan, Dapeng & Zhang, Zhiyong 2013, 'Theoretical and experimental determination of bandwidth for a two-axis fast steering mirror', Optik - International Journal for Light and Electron Optics, vol. 124, no. 16, pp. 2443–2449.

Maeda, Yutaka & Tada, Toshiki 2003, 'FPGA implementation of a pulse density neural network with learning ability using simultaneous perturbation', IEEE transactions on neural networks, vol. 14, no. 3, pp. 688–695.

Mahdieh, MH & Pournoury, M 2010, 'Atmospheric turbulence and numerical evaluation of bit error rate (BER) in free-space communication', Optics and Laser Technology, vol. 42, no. 1, pp. 55–60.

Mahdy, A & Deogun, JS 2004, 'Wireless optical communications: a survey', Proceedings of IEEE Wireless Communications and Networking Conference, WCNC, vol. 4, pp. 2399–2404.

Majumdar, Arun K; Eaton, Frank D; Jensen, Micheal L; Kyrazis, Demos T; Schumm, Bryce; Dierking, Mathew P; Dierking, Marjorie P; Shoemake, Marjorie A; Dexheimer, Dari & Ricklin, Jennifer C 2006, 'Atmospheric turbulence measurements over desert site using ground-based instruments, kite/tethered – blimp platform and aircraft relevant to optical communications and imaging systems: preliminary results', Proceedings of SPIE 6304, San Diego, California, USA. pp. 1–12.

Majumdar, Arun K & Ricklin, Jennifer C 2008, Free-Space Laser Communications: Principles and Advances: Optical and Fiber Communications Reports, Springer, USA.

Malik, Manvi & Majumder, S 2013, 'An integrated computer vision based approach for driving assistance to enhance visibility in all weather conditions', Proceedings of 1st International and 16th National conference on mechanisms, Roorkee, pp. 771–779.

Malm, William C 1999, Introduction to Visibility, Colarado State University, USA.

Matera, F & Settembre, M 2000, 'Role of Q-factor and of time jitter in the performance evaluation of optically amplified transmission systems', IEEE Journal of Selected Topics in Quantum Electronics, vol. 6, pp. 308–316.

Matsumoto, M; Kazaura, K; Dat, P; Shah, A; Omae, K; Suzuki, T; Wakamori, K; Higashino, T; Tsukamoto, K & Komaki, S 2008, 'An alternative access technology for next generation networks based on full-optical wireless communication links', Proceedings of First ITU-T Kaleidoscope Academic Conference on Innovations in NGN: Future Network and Services, Geneva, pp. 221–228.

McGuire, Patrick C; Langlois, Maud P; Lloyd-Hart, Michael; Rhoadarmer, Troy A & Angel J, Roger P 2002, 'Measurement of Atmospheric turbulence with a shack-hartmann wavefront sensor at the new MMTs prime focus', Proceedings of SPIE 4007, Adaptive Optical Systems Technology, Munich, Germany, pp. 713–721.

McGuire, Patrick C; Sandler, David G; Lloyd-Hart, Micheal & Rhoadarmer, Troy A 1991, 'Adaptive Optics: Neural network wavefront sensing, reconstruction and prediction', Proceedings of the 194th W. E. Heraeus Seminar, Bad Honnef, Germany, pp. 97–138.

Meyer-Base, A; Watzel, R; Meyer-Base, U & Foo, S 2003, 'A parallel CORDIC architecture dedicated to compute the Gaussian potential function in neural networks', Engineering Applications of Artificial Intelligence, vol. 16, no. 7–8, pp. 595–605.

Minitab Tutorials for Design and Analysis of Experiments. Available from: http://www.calpoly.edu/~pan/teaching/MinitabDOETutorial.pdf

Mndewa, Mzee S; Huang, Dexiu & Yuan, Xiuhua 2008, 'A survey of atmospheric turbulence on laser propagation', Asian Journal of Information Technology, vol. 7, no. 7, pp. 307–312.

Monin, AS & Obukhov, AM 1954, 'Basic Law of Turbulent Mixing Near the Ground', The Journal of Geophysical Institute of Slovak Academy of Sciences, USSAR, vol. 24, no. 151, pp. 163–187.

Monmasson, Eric & Cirstea, Marcian N 2007, 'FPGA Design Methodology for Industrial Control Systems – A review', IEEE Transactions of Industrial Electronics, vol. 54, no. 4, pp. 1824–1842.

Mudge, Kerry A; Silva, Dilusha; K.K.M.B., Clare, Bradley A; Grant, Kenneth J & Nener, Brett D 2011, 'Scintillation Index of the free space optical channel phase screen modeling and experimental results', Proceedings of the International Conference on Space Optical Systems and Applications, Santa Monica, CA, pp. 403–409.

Muthiah, D & Bazil Raj, A 2012, 'Implementation of High-speed LFSR Design with Parallel Architectures', IEEE Int. Conf., Computing, Communication and Applications, Dindugal, pp. 1–6.

Muthuramalingam, A; Himavathi, S & Srinivasan, E 2008, 'Neural implementation using FPGA: Issues and Applications', International Journal of Information and Communication Engineering, vol. 4, no. 6, pp. 396–402.

NABL 2000, Guidelines for estimation and expression of uncertainty in measurement, NABL, vol. 141, no. 02, pp. 1–70.

Nadeem, Farukh & Leitgeb, Erich 2010, 'Dense Maritime fog attenuation prediction from measured visibility data', Radio Engineering, vol. 19, no. 2, pp. 223–227.

Nagatsuma, Tadao 2014, 'Breakthroughs in photonics 2013: THz communications based on photonics', IEEE Photonics journal, (doi:10.1109/JPHOT.2014.2309643), vol. 6, no. 2, 0701505.

Navidpour, Mohammad S; Uysal, Murat & Li, Jing 2004, 'BER performance on MIMO free-space optical links', Proceedings of IEEE 60th Vehicular Technology Conference, pp. 3378–3382.

Nhivekar, GS & Mudholker, RR 2011, 'Data logger and remote monitoring system for multiple parameter measurement applications', Journal of Electrical and Electronics Engineering, vol. 4, no. 1, pp. 139–142.

Ni, W; Miyamoto, Y; Wakamori, K; Kazaura, K; Matsumoto, M; Higashino, T; Tsukamoto, K & Komaki, S 2009, 'Experimental study of atmospheric turbulence effects on RoFSO communication systems', PIERS Online, vol. 5, no. 1, pp. 65–70.

Nistazakis, HE; Tsiftsis, TA & Tombras, GS 2009, 'Performance analysis of free-space optical communication systems over atmospheric turbulence channels', IET Communications, vol. 3, pp. 1402–1409.

Oehlert, Gary W 2000, A First Course in Design and Analysis of Experiments, W. H. Freeman, Minneapolis.

Oh, Eun; Ricklin, Jennifer; Eaton, Frank; Gilbreath, Charmaine; Doss-Hammel, Steve; Moore, Chris; Murphy, James; Oh, Yeonju Han & Stell Mena 2004a, 'Estimating Optical Turbulence using the PAMELA Model', Proceedings of SPIE, 5550, Free-Space Laser Communications IV, Bellingham, WA, pp. 256–266.

Oh, ES; Ricklin, Jennifer C; Gilbreath, GC; Vallestero, Neil; J & Eaton Frank, D 2004b, 'Optical Turbulence Model for Laser Propagation and Imaging Applications', Proceedings of SPIE 5160, Free-Space Laser Communication and Active Illumination III, San Diego, California, USA, pp. 25–32.

Ohteru, S & Takachio, N 1999, 'Optical signal quality monitor using direct Qfactor measurement', IEEE Photonics Technology Letters, vol. 11, pp. 1307–1309.

Omondi, Amos R & Rajapakse, Jagath C 2006, FPGA Implementation of Neural Networks, Springer, Netherlands.

Pasupathi, T et al. 2014, 'FPGA Implementation of Adaptive Integrated Spiking Neural Network for Efficient Image Recognition System', ICT ACademy of Tamilnadu Journal on Image and Video Processing, vol. 4, iss. 4, pp. 848–852.

Paterson, C, Munro, I & Dainty, JC 2000, 'A low cost adaptive optics system using a membrane mirror', Optics Express, vol. 6, no. 9, pp. 175–185.

Paulsen, Uwe S; Mortensen, Niels G & Hansen, Carsten, J 2007, 'Field calibration of cup anemometers', Proceedings of European Wind Energy Conference & Exhibition, Europe, pp. 1–8.

Pederson, D & Solgaard, O 2000, 'Free space communication link using a grating light modulator', Sensors and Actuators, vol. 83, no. 1-3, pp. 6–10.

Pedireddi, Latsa Babu & Srinivasan, Balaji 2010, 'Characterization of atmospheric turbulence effects and their mitigation using wavelet-based signal processing', IEEE Transactions on Communications, vol. 58, no. 6, pp. 1795–1802.

Pelegri-Sebastia, J; Garcia-Breijo, E; Ibanez, J; Sogorb, T; Laguarda-Miro, N & Garrigues, J 2012, 'Low-cost Capacitive Humidity Sensor for Application within Flexible RFID Labels based on Microcontroller Systems', IEEE Transactions on Instrumentation and Measurement, vol. 61, no. 2, pp. 545–553.

Pena, Fernando Lopez & Duro, Richard J 2003, 'A Virtual Instrument for Automatic Anemometer Calibration with ANN based Supervision', IEEE Transactions on Instrumentation and Measurement, vol. 52, no. 3, pp. 654–661.

Peng, Yufeng; Liu, Li; Egorov, VS; Cheng, Zuhai & Zhang, Minggao 2008, 'Propagation offset characteristics of annular laser beams from confocal unstable resonators through the natural atmosphere', Optics Communications, vol. 281, no. 4, pp. 705–717.

Perez-Arancibia, Nester O; Gibson, James S & Tsao, Tsu-Chin 2012, 'Observer based intensity–feedback control for laser beam pointing and tracking', IEEE transactions on control systems technology, vol. 20, no. 1, pp. 31–47.

Perez, Joaquin; Zvanovec, Stanislav; Ghassemlooy, Zahib & Popoola, Waisu O 2014, 'Experimental characterization and mitigation of turbulence induced signal fades within an ad hoc FSO network', Optics Express, vol. 22, no. 3, pp. 3208–3218.

Perreault, JA & Bifano, TG 2004, 'High resolution parallel processor control of an optical wavefront', Proceedings of Control of Precision Systems - Digital Control System Architectues, pp. 1–6. Available from: http://www.aspe.net/publications/Spring_2004/04SP%20Extended%20Abstracts/1391-Bifano.PDF

Planchon, Thomas A; Mercere, Pascal; Cheriaux, Gilles & Chambaret, Jean-Paul 2002, 'Off-axis aberration compensation of focusing with spherical mirrors using deformable mirrors', Optics communications, vol. 216, no. 1-3, pp. 25–31.

Plett, ML; Rush, DW; Barbier, PR & Polak-Dingels, P 1999, 'Blind optimization of optical power into a single mode optical fiber using a MEMS deformable mirror', Proceedings of 2nd International Workshop on Adaptive Optics for Industry and Medicine, England, pp. 340–345.

Popoola, WO; Ghassemlooy, Z; Lee, CG & Boucouvalas, AC 2010, 'Scintillation effect on intensity modulated laser communication systems – a laboratory demonstration', Optics and Laser Technology, vol. 42, no. 4, pp. 682–692.

Prokes, Ales 2009, 'Atmospheric Effects on Availability of Free Space Optics System', Optical Engineering, vol. 48, no. 6, pp. 1–10.

Qingfang, Zhang; Hengkai, Zhao & Zhiyong, Fu 2011, 'Millimeter wave propagation characteristics trough turbulent atmosphere near the ground', Proceedings of 7th International Conference on Wireless Communications, Networking and Mobile Computing (WiCOM), Wuhan, pp. 1–4.

Rajbhandari, S; Perez, J; Le-Minh, H & Ghassemlooy, Z 2011, 'A Fast Ethernet FSO link Performance Under the Fog Controlled Environment', Proceedings of 37th European Conference and Exhibition on Optical Communication (ECOC), Geneva, pp. 1–3.

Rakotondrainibe, L; Kokar, Y; Zaharia, G & El-Zein, G 2009, '60 GHz high data rate wireless communication system', Proceedings of IEEE 69th Vehicular Technology Conference, Barcelona, pp. 1–5.

Randall, D 2004, An Introduction to atmospheric modeling, Colorado State University, USA.

Recolons, Jaume; Andrews, Larry C & Philips, Ronald L 2007, 'Analysis of Beam Wander Effects for a Horizontal-Path Propagating Gaussian-Bam Wave: Focused Beam Case', Optical Engineering, vol. 46, no. 8, pp. 1–11.

Rembe, Christian & Muller, Richard S 2002, 'Measurement system for full-three dimensional motion characterization of MEMS', Journal of Microelectromechanical systems, vol. 11, no. 5, pp. 479–488.

Ronny, ATY 2012, 'Link Performance Analysis of a Ship-to-Ship Laser Communication System', Ph.D. thesis, Naval Postgraduate School, California.

Saari, JI; Krause, MM; Walsh, BR & Kambhampati, P 2013, 'Terahertz bandwidth all-optical modulation and logic using multiexcitons in semiconductor nanocrystals', Nano letters, vol. 13, no. 2, pp. 722–727.

Sadot, D & Kopeika, NS 1992, 'Forecasting optical turbulence strength on the basis of macroscale meterology and aerosols: models and validation', Optical Engineering, vol. 31, no. 2, pp. 200–212.

Safari, Majid & Uysal, Murat 2008, 'Do we really need OSTBCs for Free space optical communication with direct detection?', IEEE transactions on wireless communications, vol. 7, no. 11, pp. 4445–4448.

Sandalidis, HG 2011, 'Coded free-space optical links over strong turbulence and misalignment fading channels', IEEE Transactions on Communications, vol. 59, no. 3, pp. 669–674.

Santiago, Freddie; Wilcox, Christopher; Chang, Mark; Font, Carlos O; Roura, Erick & Restaino, Sergio 2005, 'Low altitude horizontal scintillation measurements of atmospheric turbulence over the sea: Experimental results', Proceedings of SPIE 6014, Active and Passive Optical Components for WDM Communications, Boston, pp. 321–329.

Saunter, CD; Love, GD; Johns, M & Holmes, J 2006, 'FPGA technology for high speed low cost adaptive optics' Proceedings of SPIE 6018, 5th International Workshop on Adaptive Optics for Industry and Medicine, China, pp. 60181G.

Schneider, Thomas; Wiatrek, Andrzej; Preubler, Stefan; Grigat, Michael & Braun, Ralf-Peter 2012, 'Link budget analysis for terahertz fixed wireless links', IEEE transactions on terahertz science and technology', vol. 2, no. 2, pp. 250–256.

Scintec Corporation 2012, Laser Scintillometer SLS20 Datasheet. Available from: http://www.scintec.com/english/CustomUpload/374035703400370035603690350034003650360037303560369037003720364O/SLS20_2012.pdf

Sensirion 2011, SHT1X Datasheet. Available from: http://www.sensirion.com/fileadmin/user_upload/customers/sensirion/Dokumente/Humidity/Sensirion_Humidity_SHT1xDatasheet_V5.pdf

Sergeyev, Aleksandr & Roggemann, Michael 2013, 'Ansioplanatic Studies and Fried Parameter Estimation via Multi-Channel Laser Communication', IEEE Aerospace Conference, Big Sky, MT, pp. 1–9.

Sevincer, Abdullah; Bilgi, Mehmet; Yuksel, Murat & Pala, Nezih 2010, 'Prototyping multi-transceiver free space optical communication structures', Proceedings of the International Conference on Communications, Cape Town, South Africa, pp. 1–5.

Shake, I; Takara, H; Uchiyama, K & Yamabayashi, Y 2001, 'Quality monitoring of optical signals influenced by chromatic dispersion in a transmission fiber using averaged Q-factor evaluation', IEEE Photonics Technology Letters, vol. 13, pp. 385–387.

Sharma, Manish & Chadha, D 2006, 'Maximum likelihood decoding of low density parity check coded subcarrier modulation in optical wireless systems', Proceedings of Annual IEEE India Conference, New Delhi, pp. 1–6.

Shen, Jing-Chung; Jywe, Wen-Yuh; Chiang, Huan-Keng & Shu, Yu-Ling 2007, 'Precision tracking control of a piezoelectric actuated system', Proceedings of 15th Mediterranean Conference on Control and Automation, Athens, pp. 1–6.

Si, Congfang & Zhang, Yixin 2013, 'Beam wander of quantization beam in a non-Kolmogorov turbulent atmosphere', Optik – International Journal for Light and Electron Optics, vol. 124, no. 12, pp. 1175–1178.

Sikora, A & Groza, VF 2005, 'Coexistence of IEEE802. 15.4 with other Systems in the 2.4 GHz-ISM-Band', Proceedings of IEEE Conference on Instrumentation and Measurement Technology, Ottawa, Ont, pp. 1786–1791.

Silva, Vinicius NH; Barbero, Andres PL & Ribeiro, Ricardo M 2011, 'A new triangulation-like technique for the evaluation of the refractive index structure constant (C_n^2) in Free-space optical links', Journal of Lightwave Technology, vol. 29, no. 24, pp. 3603–3610.

Silva, VNH; Barbero, APL & Ribeiro, RM 2011, 'A New Triangular-like technique for the evaluation of the refractive Index structure constant (Cn2) in Free-Space Optical Links', Journal of Lightwave Technology, vol. 29, no. 24, pp. 3603–3609.

Sinefeld, D; Shayovitz, D; Golani, O & Marom, DM 2013, 'Adaptive Rate and Bandwidth WDM Optical Sampling Pulse Streams with LCoS-based Photonic Spectral Processor', Proceedings of Optical Fiber Communication Conference, Anaheim, CA, pp. 1–3.

Singh, Chaturi; John, Joseph; Singh, YN & Tripathi, KK 2004, 'Indoor optical wireless systems: design challenges, mitigating techniques and future prospects', IETE Technical Review, vol. 21, no. 2, pp. 101–117.

Sitterle, JJ; Podany, ME; Galm, JM; Claspy, PC & Merat, FL 1988, 'A Meteorological Data Acquisition System for Atmospheric Propagation Studies', IEEE Transactions on Instrumentation and Measurement, vol. 37, no. 3, pp. 444–448.

Skoda, P; Lipic, T; Srp, A; Medved Rogina, B; Skala, K & Vajda, F 2011, 'Implementation Framework for Artificial Neural Networks on FPGA', Proceedings of 34th International Convention MIPRO, Opatija, pp. 274–278.

Smith, Scott T 2006, MATLAB Advanced GUI Development, Dog Ear Publishing, USA.

Sofiya, J et al. 2015, 'Certain Investigations on Atmospheric Turbulence Strength (C_n^2) prediction models for FSO Applications', International Journal for Technological Research in Engineering, vol. 2, iss. 8, pp. 1579–1582.

Soibel, A; Wright, MW; Farr, WH; Keo, SA; Hill, CJ; Yang, RQ & Liu, HC 2009, 'Midinfrared Interband Cascade Laser for Free Space Optical Communication', IEEE Photonics Technology Letters, vol. 22, no. 2, pp. 121–123.

Song, Qing; Wu, Yilei & Soh, Yeng Chai 2008, 'Robust adaptive gradient-descent training algorithm for recurrent neural networks in discrete time domain', IEEE transactions on neural networks, vol. 19, no. 11, pp. 1841–1853.

Sprangle, Phillip; Ting, Antonio; Penano, Joseph; Fischer, Richard & Hafizi, Bahman 2009, 'Incoherent combining and atmospheric propagation of high-power fiber lasers for directed-energy applications', IEEE journal of Quantum Electronics, vol. 45, no. 2, pp. 138–148.

Suite, MR; Burris, HR; Moore, CI; Vilcheck, MJ; Mahon, R; Carmen Jackson; Stell, MF; Davis, MA; Rabinovich, WS; Scharpf, WJ; Reed, AE & Gilbreath, GC 2004, 'Fast Steering Mirror Implementation for Reduction of Focal – Spot Wander in a Long Distance Free-Space Communication Link', Proceedings of SPIE 5160, Free Space Laser Communication and Active illumination, San Diego, California, USA, pp. 439–446.

Tang, X; Ghassemlooy, Z; Rajbhandari, S; Popoola, WO; Lee, CG; Leitgeb, E & Ahmadi, V 2010, 'Free space optical communication employing polarization shift keying coherent modulation in atmospheric turbulence channel', Proceedings of 7th International Symposium on Communication Systems Networks and Digital Signal Processing, Newcastle upon Tyne, pp. 615–620.

Taylor, FJ; Mohan, P; Joseph, Philip & Pries, TH 1977, 'An All Digital Automated Wind Measurement System', IEEE Transactions on Instrumentation and Measurement, vol. 26, no. 2, pp. 175–180.

Texas Instruments, 2013, ADC12D1000. Available from: http://www.ti.com/lit/ds/symlink/adc12d1000.pdf

Tisan, Alin; Oniga, Stefan; Mic, Daniel & Buchman, Attila 2009, 'Digital Implementation of the Sigmoid Function for FPGA Circuits', ACTA Technica napocensis, vol. 50, no. 2, pp. 15–20.

Tisan, Alin & Cirstea, Marcian 2013, 'SOM Neural Network Design – a new simulink library based approach targeting FPGA Implementation', Mathematics and Computers in Simulation, vol. 91, pp. 134–149.

Toyoshima, M; Takahashi, Nobuhiro; Jono, Takashi; Yamawaki, Toshihiko, Nakagawa, Keizo & Yamamoto, Akio 2001, 'Mutual alignment errors due to the variation of wave-front aberrations in a free-space laser communication link', Optical Express, vol. 9, no. 11, pp. 592–602.

Trisno, Sugianto; Ho, Tzung-Hsien; Milner, Stuart D & Davis Christopher C 2004, 'Theoretical and experimental characterization of omnidirectional optical links for free space optical communications', Proceedings of IEEE military communications conference, pp. 1151–1157.

Tsiftsis, Theodoros A; Sandalidis, Harilaos G; Karagiannidis, George K & Sagias, Nikos C 2006, 'Multihop free space optical communications over strong turbulence channels', Proceedings of IEEE International Conference on Communications, Istanbul, pp. 2755–2759.

Tsiftsis, Theodoros A; Sandalidis, Harilaos G; Karagiannidis, George K & Uysal, Murat 2009, 'Optical wireless links with spatial diversity over strong turbulence channels', IEEE transcations on wireless communications, vol. 8, no. 2, pp. 951–957.

Tsou, Haiping & Yan, Tsun-Yee 1999, 'The extended-image tracking technique based on the maximum likelihood estimation', pp. 1–8. Available from: http://trs-new.jpl.nasa.gov/dspace/handle/2014/17675/1/99-1109.pdf

Tunick, Arnold 2007a, 'Statistical Analysis of Measured Free-Space Laser Signal Intensity over a 2.33 km Optical Path', Optics Express, vol. 15, no. 21, pp. 14115–14122.

Tunick, Arnold 2007b, 'Statistical analysis of optical turbulence intensity over a 2.33 km propagation path', Optics Express, vol. 15, no. 7, pp. 3619–3628.

Tunick, Arnold 2008, 'Optical turbulence parameters characterized via optical measurements over a 2.33 km free-space laser path', Optics Express, vol. 16, no. 19, pp. 14645–14654.

Tunick, Arnold; Tikhonov, Nikolay; Vorontsov, Mikhail & Carhart, Gary 2005, 'Characterization of Optical Turbulence (C_n^2) data measured at the ARLA_LOT Facility', (ARL–MR–625).

Usman, A; Olaore, KO & Ismaila, GS 2013, Estimating visibility using some meterological data at sokoto, Nigeria', International journal of basic and applied science, vol. 1, no. 1, pp. 810–815.

Valle, Maurizio & Diotalevi, Francesco 2004, 'A dedicated very low power analog VLSI architecture for smart adaptive systems', Applied soft computing, vol. 4, no. 3, pp. 206–226.

Vasumathi, B & Moorthi, S 2012, 'Implementation of hybrid ANN–PSO algorithm on FPGA for harmonic estimation', Engineering Applications of Artificial Intelligence, vol. 25, no. 3, pp. 476–483.

Vladutescu, Daniela Viviana; Wu; Gross, Barry M; Moshary, Fred; Ahmed, Samir A; Blake, Reginald A & Razani, Mohammad 2012, 'Remote Sensing Instruments used for Measurement and Model Validation of Optical Parameters of Atmospheric Aerosols', IEEE Transactions on Instrumentation and Measurement, vol. 61, no. 6, pp. 1773–1746.

VTI Technologies 2007, SCP 1000 Datasheet. Available from: https://www.sparkfun.com/datasheets/Components/SCP1000-D01.pdf

Vu, Bach T; Dang, Ngoc T; Thang, Truong C & Pham, Anh T 2013, 'Bit Error Rate Analysis of Rectangular QAM/FSO Systems Using an APD Receiver Over Atmospheric Turbulence Channels', Journal of Optical Communications and Networking, vol. 5, no. 5, pp. 437–446.

Wakamori, K; Kazaura, K & Oka, I 2007, 'Experiment on Regional Broadband Network Using Free-Space-Optical Communication Systems', Journal of Lightwave Technology, vol. 25, no. 11, pp. 3265–3273.

Wan, Lingyu; Liu, Liren & Sun, Jianfeng 2010, 'On-ground simulation of optical links for free-space laser communications', Optik – International Journal for Light and Electron Optics, vol. 121, no. 3, pp. 263–267.

Wang, Jin; Kahn, Joseph M & Lau, Kam Y 2002, 'Minimization of acquisition time in short-range free-space optical communication', Applied Optics, vol. 41, no. 36, pp. 7592–7602.

Wekesa, DW; Kamau, JN & Mutuku, JN 2013, 'Calibrated data logging instrumentation system for wind speed and direction measurements', Basic Research Journal of Engineering Innovation, vol. 1, no. 3, pp. 53–57.

Willebrand, H & Ghuman, BS 2002, 'Free Space Optics: Enabling Optical Connectivity in Today's Networks, Sams Publishing, London.

Witzel, John 2008, 'Spelunkers, how deep are you? [my favorite experiment], IEEE Instrumentation & Measurement Magazine, vol. 11, no. 3, pp. 41–43.

Wood, Robert J; Steltz, Erik & Fearing, Ronald S 2005, 'Nonlinear performance limits for high energy density piezo electric bending actuators', Proceedings of IEEE International Conference on Robotics and Automation, Barcelona, Spain, pp. 3633–3640.

Wright, NG & Easom, GJ 2003, 'Non-linear k-ε turbulence model results for flow over a building at full scale', Applied Mathematical modeling, vol. 27, no. 12, pp. 1013–1033.

Wu, H & Kavehrad, M 2007, 'Availability evaluation of ground-to-air hybrid FSO/RF links', International Journal of Wireless Information Networks, vol. 14, pp. 33–45.

Wu, Hanling; Yan, Haixing & Li, Xinyang 2010, 'Modal correction for fiber-coupling efficiency in free-space optical communication systems through atmospheric turbulence', Optik – International Journal for Light and Electron Optics, vol. 121, no. 19, pp. 1789–1793.

Xie, Wen-Fang; Fu, Jun; Yao, Han & Su, CY 2009, 'Neural Network-based Adaptive Control of Piezoelectric Actuators with Unknown Hysteresis', International Journal of Adaptive Control and Signal Processing, vol. 23, no. 1, pp. 30–54.

Yague, C; Maqueda, G & Rees, JM 2001, 'Characteristics of turbulence in the lower atmosphere at Halley IV station, Antartica', Dyanmics of Atmospheres and Oceans, vol. 34, no. 2-4, pp. 205–223.

Yahaya, S & Frangi, JP 2004, 'Cup Anemometer response to the wind turbulence – measurement of the horizontal wind variance', Annales Geophysicae, vol. 22, pp. 3363–3374.

Yan, Peng; Sluss Jr, James J; Refai, Hazem H & Lopresti, Peter G 2005, 'An initial study of mobile Adhoc networks with free space optical capabilities', Proceedings of 24th Digital Avionics Systems Conference, pp. 1.D.3–11-10.

Yang, Tsung-Yi; Gourlay, James & Walker, Andrew C 2002, 'Adaptive Alignment Packaging for 2-D Arrays of Free – Space Optical – Interconnected Optoelectronic Systems', IEEE Transactions on Advanced Packaging, vol. 25, no. 1, pp. 54–63.

Yellow Fourier Technologies, 2006, Eye Diagram Analyzer. Available from: http://yellowfourier.com/eyedia.html

Yildirum, Sahin 2006, 'A proposed neural internal model control for robot manipulators', Journal of Scientific and Industrial Research, vol. 65, pp. 713–720.

Yitzhaky, Yitzhak; Dror, Itai & Kopeika, Norman S 1997, 'Restoration of atmospherically blurred images according to weather-predicted atmospheric modulation transfer functions', Optical Engineering, vol. 36, no. 11, pp. 3064–3072.

Yoshida, Koichi; Tanaka, Kuniaki; Katayama, Kazunori; Tsujimura, Takeshi & Azuma, Yuji 2009, 'Collimator Focus Adjustment for Free Space Optics System using Single-Mode Optical Fibers', Proceedings of ICROS – SICE International Joint Conference, Fukuoka, pp. 1338–1341.

Yu, S; Alici, G; Shirinzadeh, B & Smith, J 2005, 'Sliding Mode Control of a Piezo electric Actuator with Neural Network Compensating Rate-Dependent Hysteresis', Proceedings of International Conference on Robotics and Automation, pp. 3641–3645.

Yuksel, Murat; Akella, Jayasri; Kalyanaraman, Shivkumar & Dutta, Partha 2009, 'Free-Space Optical Mobile Adhoc Networks: Auto – configurable building blocks', Wireless Networks, vol. 15, no. 3,pp. 295–312.

Zabidi, SA; Khateeb, WA; Islam, MR & Naji, AW 2010, 'The effect of weather on free space optics communication (FSO) under tropical weather conditions and a proposed setup for measurement', Proceedings of International Conference on Computer and Communication Engineering (ICCCE), Kuala Lumpur, pp. 1–5.

Zamek, Steve & Yitzhaky, Yitzhak 2006, 'Turbulence strength estimation from an arbitrary set of atmospherically degraded images', Journal of Optical Society of America A, vol. 23, no. 12, pp. 3106–3113.

Zhao, Zhijun; Liao, Rui, Lyke, Stephen D & Roggemann, Micheal C 2010, 'Reed-Solomon coding for Free-space optical communications through turbulent atmosphere', Proceedings of IEEE Aerospace Conference, Big Sky, MT, pp. 1–12.

Zhao, Zhijun; Liao, Rui; Lyke, Stephen D & Roggemann, Michael C 2010, 'Direct detection free-space optical communications through atmospheric turbulence', Proceedings of Aerospace conference, Big Sky, MT, pp. 1–9.

Zhen, Hu; Zhengxum, Song; Shoufeng, Tong; Xin, Zhao; Hongfei, Song & Huilin, Jiang 2009, 'Modeling of fine tracking sensor for free space laser communication systems', Proceedings of symposium on Photonics and Optoelectronics, Wuhan, pp. 1–4.

Zheng, Chongwei 2014, 'Propagation of elliptical gaussian beam passing through apertured paraxial optical systems', Optik – International Journal for Light and Electron Optics, vol. 125, no. 1, pp. 264–267.

Zilberman, Arkadi; Golbraikh, Ephim & Kopeika, Norman S 2010, 'Some limitations on optical communication reliability through kolomogorov and non-kolomogorov turbulence', Optics communications, vol. 283, pp. 1229–1235.

Zocchi, Fabio E 2005, 'A simple analytical model of adaptive optics for direct detection free-space optical communication', Optics Communications, vol. 248, pp. 359–374.

Zvanovec, S; Perez, J; Ghasemloy, Z; Rajbhandari, S & Libich, J 2013, 'Route diversity analyses for fre-space optical wireles links within turbulent scenarios', Optics Express, vol. 21, iss. 6, pp. 7641–7650.

Index

absolute humidity 57
absorption 2, 6, 55, 94, 115, 145
abstraction module 126
accelerometer 96, 119
accurate prediction 47, 60, 81, 92
acquisition process 14
activation functions 108
A/D 31, 97, 98, 120, 121, 128, 129
adaptive alignment 96
adaptive control 155
adaptive optics 94–96, 119
adaptive technique 47
aerosol 6, 15, 52, 55, 60
aerosol scattering 6, 56
aerospace communications 3
aircraft to aircraft 150
altitude 8, 13, 16, 34, 36, 44, 47, 49, 52–54, 58, 81, 93, 96, 97, 115, 117, 120, 146
ambient conditions 3
analysis of variance 47, 95
angle of arrival 15, 118
ANOVA 9, 47, 62, 65, 92, 95, 109, 122, 146
aperture averaging 14
architecture 8, 14, 18, 20, 22, 26, 29, 30, 45, 117, 118, 123, 125, 128, 131, 134, 135, 146
arithmetic operations 122, 134
artificial neural network 8, 15, 93, 95, 117, 118
ASIC 24
asynchronous transfer mode 160, 178
atmosphere 1, 4, 6, 8, 37, 39, 41, 43, 47, 48, 50, 54, 56, 57, 60, 68–70, 73, 75, 77, 80, 83, 87, 89, 94, 117, 146
atmosphere turbulence 4
atmospheric absorption 6
atmospheric attenuation 8–10, 13, 48–51, 53, 56, 60, 62, 67, 69, 70, 72, 75, 77, 79, 149, 152, 154, 156, 158, 162, 164, 173, 175, 176, 179
atmospheric attenuation coefficient 49
atmospheric chamber 50, 52
atmospheric changes 13, 49, 117
atmospheric channel 1, 2, 5, 6
atmospheric effects 10, 149, 153, 156
atmospheric particles 2
atmospheric refractive index 5, 35, 94
attenuation 4, 8, 47–52, 54–57, 60, 62, 63, 66, 67, 69–73, 75–78, 92, 149, 152, 154, 156, 162, 176, 179
auto-tracking 94
average transmit optical power 165
azimuth and elevation distance 122

back-propagation 105, 119, 125
backhaul traffic 13
bandwidth capability 2
Bataille's model 57, 72, 74, 77, 79
battlefield 1, 93
Bayesian approach 154
beacon beam 160
beam centroid information 47, 161, 176
beam fluctuation 8, 95
beam motion 93
beam spot centroid motion 114, 144
beam stabilization 7, 109, 112, 113, 119, 122, 136, 140, 142, 143, 147
beam stabilizing system 162
beam steering 15, 94–98, 103, 113–115, 118, 119, 121, 145, 146
beam steering control 118, 155, 168, 169, 177
beam steering controller 95, 118
beam surface 152
beam wander 51, 59, 117, 136
beam wandering compensation 149, 155, 163, 166
beam wandering model 59
Beer–Lambert law 49
binary phase shift keying 155, 178
bipolar linear activation function 131
bit error rate 5, 94, 118, 149, 154, 157, 178
bit time 161, 163
BKB model 51
blurred image 52
bottleneck 2, 4
broadband internet 151
broadband telecommunication 2
broadcasting services 94, 151
building sway 94, 117, 119

calibration 9, 14, 15, 19, 21, 32, 52, 95, 100–102, 115, 119
calibration result 100

capacity 1–4, 47
capacity limiting factors 48
Cartesian coordinate 59
centroid stability 149
channel behavior 153
channel behaviour 154, 175
channel capacity 152
channel control register 128
channel distance 150
channel select 128
circular shift register 23, 24, 133
clock manager unit 128
closed loop control 136, 140–142, 145
cloud cover 34, 58
coefficient of determination 33, 38, 39, 42, 44, 47, 62, 65, 92, 109, 116
coherent radius 153
combined uncertainty 34
communication frame 14, 26, 29, 30, 135
communication technologies 3
compression 162
computational latency 126
confirmatory test 10, 92, 116
control engine 22–24, 26–28, 126
CORDIC 126
corrected wavefront 155
correlation coefficients 8, 13, 32, 45
correlation level 139
cubic equation 47, 61
cup anemometers 15
cut-off frequency 161

D-FFs 129
data conversion 9, 14
data logging 29, 31, 35, 36, 45, 47, 49, 53, 54
data routing ring circuit 132
data transmission 7, 8, 10, 47, 118
daylong operation 172
daywise data 54
deep fades 153
deep signal fading 95, 144
deep space 1, 94, 96, 119
deep space applications 1
deep space optical communications 96, 119
defense applications 4, 94
defense communications 7
deformable mirror 179
degradation 5, 95, 149, 153, 155, 166, 179
degree of freedom 34, 96
design of experiment 104
detector plane 7, 10, 113, 117, 140, 143
different modulation schemes 178
digital architectures 13, 18
digital clock manager 128
digital subscriber loop 4
digitization 128, 135
directionality 3, 5
discrete wavelet transform 155
displacement error 96, 118, 122
disturbance rejection 114, 140, 145
diurnal pattern 61
divergence angle 13

earth observation satellite 49
earthquakes 94, 117
eddies of air 48
eddy viscosity 35
effective scintillation index 144
electro-optics 4
electromagnetic interference 175
EMC 150
EMI 93, 150
empirical models 50
enhancement in channel modeling 179
ensemble 18, 38, 40–43, 60, 145
environmental conditions 7–9, 13, 14, 36, 44, 45, 54
error histogram 137
estimation accuracy 9
execution command 23
expanded uncertainty 34
experiment parameters 55
experimental design 104
experimental test-bed 14, 159
eye diagram 149, 157, 161, 162, 170, 173
eye-opening 163, 174
eye safe 6

fading 96, 150, 152, 155, 180
fast-Ethernet 155
fast steering mirror 93, 95, 119
feedforward 125, 128
field programmable gate array 8, 96, 117, 118
finite state machine 23, 27
finite state machine engine 126
five nines 2, 152
flexibility 2, 93, 124, 126
floating point arithmetic 122, 134

floating point calculations 119
forecasting 14, 15
form factors 5
forward error control 170, 180
FPGA 8, 10, 14, 17, 18, 20, 22, 29, 44, 96, 117–120, 122, 124–128, 131, 135, 146
free space optics communication 1
frequency allocations 150
full-angle divergence 156
full duplex FSOC 178
full model 104
full model equation 61
function approximation 103, 131
future optical networks 94

Gaussian distribution 157
Gaussian laser beam 155
geographical location 34, 51, 58
geographical parameters 16
geometrical interpretation 166, 167
geometrical loss 165
GMT + 5.30 34
gradient descent algorithm 105
graphical user interface 54
ground station to aircraft 150
ground stations 3

hardware interfacing 18
hidden layer 105, 124, 133
high bandwidth 4, 7, 94
high data rates 1, 7, 94, 178
Hilbert–Huang 16
histogram 113–115, 137, 144
horizontal propagation 49, 152
Hufnagel–Valley model 16, 58
hybrid FSOC/RF communication 180
hyperbolic tangent function 131
hysteresis 96, 101, 103, 119
hysteresis nonlinearity 96, 103

immunity 93, 150, 175
impulse response 122, 146, 147
indoor beam stabilization 95, 119
information capacity 1
information technology 49, 120
inhomogeneities 59
inhomogeneous turbulence 47
inner scale magnitude 129
input layer 105, 124
intelligent tool 95, 118
intensity fluctuation 6, 14
intensity noise 48
inter-satellite 3, 94
inter-satellite link 3
interaction equation 61
interaction model 104
interfacing technique 9
interference 4, 5, 142
interstellar communication 7
irradiance 5, 18, 35, 48, 51, 59, 145
Itai Dror's model 57

jitter 97, 121, 160, 176
JK FF 20

Kalman filter 95, 119
Koschmieder law 49

large data bandwidth 150
laser communication laboratory 16, 53, 97
laser cutting 93, 116
last mile access 2
last mile access network 2
last mile connectivity 3, 150
least mean square 105
license-free 6
license-free band of operation 3, 93
light signals 1
line-of-sight 1, 156
linear equation 61
linear model 104
linear regression analysis 47
liner time invariant 126
linguistic terms 168
link availability 2, 52
link budget 52
link margin 52, 149, 159, 161, 165, 174, 178
link performance 93, 118
link range 1, 2, 8, 13, 18, 44, 49, 51, 53, 149, 150, 159, 166, 174, 178, 179
local area networks 4
local time 34, 37, 39, 41, 44, 58
local weather data 50, 53, 63, 149, 156
log-normal FSOC channel 154
long range communications 1
long-term data loss 95
look-up table 131
low altitude 9

low pass filter 161, 163
low power usage per bit 3, 93

M. Ijaz's model 56, 68, 70, 73, 75, 78
M-PAM 178
macro-tip-tilt 119
magnetic pole 19
major power loss 93, 117, 149
Mars laser communication demonstration 1
Mathcad version 35
mathematical models 47
maximum data rate 10, 49, 153, 178
maximum transmittance 156, 158, 162
mean and variance 157
measurement 8–10, 13–16, 19–29, 31–33, 36–45, 48, 50–52, 54, 58, 63, 65, 67, 68, 71, 73, 74, 76, 82, 86, 92, 100, 102, 111, 122, 135, 136, 143, 146
measurement system 13, 14, 32, 45
mechanical vibrations 95, 119, 151
meteorological parameters 9, 13, 18, 34, 36, 38, 40–43, 47, 49, 56–58, 60, 62, 63, 65, 67, 81, 83, 86, 88, 90
metropolitan area networks 4
metropolitan network 3
micro-radians 156
military 1, 7
MIMO 180
MISO 26
misty 37, 87, 89
mitigation 5, 50
model equation I 63, 65
model equation II 64
model equation III 64
model equation IV 64
model equation V 47, 62, 64, 66
models 8–10, 16, 34, 47, 50–52, 54, 56, 58, 60–62, 65–70, 72–75, 77–80, 82, 84, 85, 87, 89–92, 102, 110, 111, 119, 120, 122
modern information technology 149
modulated beam 117, 155, 156, 164
modulation techniques 152, 180
molecular scattering 6
mono-pulse arithmetic circuit 53, 99
mono-pulse 8
monsoon 41, 67
MOSI 26
M. S. Awan's model 56, 69, 71, 73, 75, 78
multi-Gb/s 3
multilayer perceptron 105, 106, 125
multimedia society 3
multipath propagation 180
multiplexer 23, 122, 128, 131, 133
multiply-accumulation unit 132
multiply-accumulator unit 134

NABL 31, 34
NBIOF 53
near earth FSOC 2
neural-controller 8, 95, 97–99, 105, 108–112, 115–118, 120, 122, 124–128, 136–140, 142, 145–147
neuron 105, 120, 131–134
next generation networks 13
nonlinear 93, 103, 119
nonlinear bipolar activation function 132
normalization 103, 122, 123, 129
NPL 31
NRZ-OOK-IM-DD 157, 158, 160, 178

observer-based feedback 95, 119
Obukhov buoyancy length 35
on-off-keying 178
open atmosphere 155
open loop decision-making 136
optical breadboard 9, 54, 102
optical collimator 161
optical communication link 155
optical fiber amplifiers 152
optical irradiance 5
optical loss 13
optical noise 153
optical power fluctuation 115, 145, 152
optical repeaters 3
optical scattering 152
optical shaker 95, 97, 119–121
optical spectrum 4
optical wireless communication 1
optical wireless link 17, 93
optoelectronic components 3, 54, 97, 140
optoelectronic position detector 51, 99
optoelectronics 4, 16, 52–54, 117
outer-scale turbulence 155
output enable 128
output layer 105, 124, 133

PAMELA model 15, 16, 34–36, 38–45, 58, 80, 85, 87, 90

parallel processor 8
Pasquill stability category 35
peak to average power ratio 5, 164
percentage deviation 65, 66, 92
percentage of absolute error 52
phase fluctuations 94
photo-phone 1
photodetector sensitivity 165
phototransistors 15
piezo-actuators 119
piezoelectric actuators 97, 98, 103, 159, 176
pipelined-parallel architecture 123
point-to-point 3, 93
pointing acquisition and tracking 8, 94
pointing error 96, 149, 156, 159, 162, 165, 174
pointing loss 13
pointing misalignment 153
polarization drift 94
polynomial regression 59
power efficiency 2
power in the bucket 117, 167
power-line communication 4
power spectral density 140, 141
PRBS 156, 159, 178
pre-processor module 128
prediction accuracy 8, 47, 56, 61, 62, 65, 67, 70, 72, 139
pressure 5, 8, 13, 14, 18, 24, 26–28, 31, 33, 35, 37–44, 54, 80, 81, 83–90, 94
presummer 39, 75
probability dispersion 34
probability of intercept 4
process-time diagram 126
process variables 128
processing gain 111
propagating laser 9, 48
pseudocode 21, 24, 25, 28
pure reflection mirror 97, 160

Q-factor 149, 155, 157–161, 163–165, 168–171, 173, 174, 178
Q-PSK 178
quadratic equation 61
quadratic model 104
quick and easy installation 3

radial displacement 51, 59
radiation class 35
radio over FSOC 180
radix-2 format 29, 36, 135
rainy 43, 70
receiver sensitivity 13
relative humidity 8, 13–15, 22–24, 31, 33, 37–44, 51, 52, 54, 56, 57, 60–62, 65, 67, 69, 73, 75, 77, 80, 81, 83, 85–90, 94
residual plots 63, 66
resolution 21, 24, 31, 34, 101, 124
response surface 47, 62, 65, 110, 112, 116, 123, 124, 137–139
response surface model 8, 10, 62, 95, 110, 117, 118, 138
RF-FSO test link 51
root mean square error 9, 47, 51
route diversity 180
RS232 communication 29, 99
Rytov method 14, 36

satellite crosslinks 150
satellite networks 150
satellite platform 3
satellite to aircraft 150
scattering 2, 6, 15, 48, 52, 55, 94, 110, 115, 145
scintillation 2, 6, 13, 17, 48, 51, 60, 94, 117, 144, 145, 147
scintillation effects 13, 144
scintillation index 18, 145
scintillometer 13, 15, 16, 18, 36–38, 40–44, 52
SCP1000-D01 14, 24
secure communications 1
semiconductor-laser inter-satellite link experiment 1
sensors 13–15, 18, 19, 32, 36, 44, 49, 54, 96, 146
serial clock 22
serial communication manager 135
serial peripheral interface 18, 26
sessions 14, 15
setpoint 32, 112
ship-to-ship communications 7
SHT11 14, 21–24
sigmoidal activation function 131
signal analysis 160
signal to noise ratio 94, 149, 157, 159
signaling format 163, 164
simplex communication 8
simulation results 9, 51, 96, 120, 132
single mode fiber 95
single-precision 122, 134

SISOCSR 132
SLC-day 34
space scientific exploration 150
spatial diversity 180
spatial domain 94
standard deviation 37
steering operation 122
steering technique 8, 116
subcarrier intensity modulation 155
successive computations 130
sum of absolute error 47, 80
summer 39
sunlight 1
surface boundary layer 34, 58
surface roughness length 35, 58
synchronization 14, 18, 23, 27
system capacity 1

Taguchi method 93
telecommunications networks 4
telescope 53, 97, 120
temperature 5, 8, 13, 18, 22–24, 31, 33, 35–44, 51, 52, 54, 56–58, 60–62, 65, 67, 70, 73, 75, 77, 80, 81, 83, 85–90, 94, 96
temporal broadening 154
terrain type 34, 58, 120
terrestrial communications 94
terrestrial FSOC 2
terrestrial last mile 93
TFSLSOC 8
threshold values 149, 178
tip-tilt mirror 119
training process 105, 108, 125
trans impedance amplifier 160
transmission bandwidth 13
transmission start sequence 23
transmittance 6, 49, 56, 60, 149, 156–158, 161–165, 174, 178
transmitting optics 53, 118, 156, 159, 160
trapezoidal distribution 137
traveling wave 59, 156
turbulence simulation box 96
turbulence strength 7–10, 13, 14, 16, 17, 34–38, 40–44, 47, 49–54, 58–60, 62, 65–67, 82, 84, 87, 89, 91, 92, 149, 154–156, 161, 165, 166, 170, 171, 173, 175, 176, 179, 180
turbulent atmospheric 17, 149, 152, 173
two wire interface 18, 22
type A & type B 31

UART 14, 29, 30, 45, 118, 127, 135, 146
ubiquitous communication 151
uncertainty 9, 14, 15, 31, 33, 34
unlicensed modulation 2

variable beam splitter 53, 97, 120, 121
vertical momentum flux 35
very strong atmospheric 149, 161, 164, 166, 171, 180
VHDL 18, 21, 25, 28, 120, 126
vibration 9, 16, 54, 97, 102, 117, 120
vibration damped 9, 16, 54, 97, 102, 120
visibility 13, 49–52, 56, 60, 67, 68, 70, 71, 73–78, 92

wave number 17
wavefront aberration 94
wavefront correction 94
wavefront distortion 7
wavefront distortion correction 179
wavefront distortions 48, 95, 152
wavefront profile 167
wavelength dependence 179
weather condition 2
weather seasons 9
weather sensors 9, 18
weather station 18, 47, 54, 60
wind shear 35, 49, 58
wind speed 8, 13, 14, 16, 19–21, 31, 33, 35, 37–44, 54, 57–62, 65, 67, 70, 73, 75, 77, 80, 81, 83, 86–90, 94
winter 37
wireless applications 2
wireless optical communication 153
wireless solution 3
wireless systems 2

Zernike model 96
z_{offset} 101, 143

Bei Fragen zur Produktsicherheit wenden Sie sich bitte an:
If you have any questions regarding product safety,
please contact:

Walter de Gruyter GmbH
Genthiner Straße 13
10785 Berlin
productsafety@degruyterbrill.com